Fundamentals of Plants Biotechnology and Pathology

About the Author

Saba Hussain, is post Graduate in Biotechnology from Dr.Bhimrao Ambedkar University, Agra, Uttar Pradesh and Submitted PhD in Biotechnology from C.M.J.University Shillong, Assam. Presently she is working as a Assitant Professor in Biotechnology Department at Patna Science College of Pharmaceutical Education & Research, Patna, Bihar. She has attented various seminars and Conferences and Presented International papers.

About the Book

Plant biotechnology is a set of techniques used to adapt plants for specific needs or opportunities. Situations that combine multiple needs and opportunities are common. For example, a single crop may be required to provide sustainable food and healthful nutrition, protection of the environment, and opportunities for jobs and income. Finding or developing suitable plants is typically a highly complex challenge. Plant biotechnologies that assist in meeting the challenge include genomics, molecular-assisted selection, and transgenic crops. These biotechnologies allow researchers to detect and map genes, discover their functions, select for specific genes in genetic resources and breeding, and transfer genes for specific traits into plants where they are needed. NIFA funds research, training, and extension for developing and using biotechnologies for food and agriculture.

Fundamentals of Plants Biotechnology and Pathology

Saba Hussain

AMIGA PRESS INC.
DELHI (INDIA)

AMIGA PRESS INC.
H.O.: 28-A, Indra Park,
Chander Nagar, Delhi-110051 (India)
Tel.: 011-22504028, +91-98736-57000
E-mail: amigapress@gmail.com
amigaeditorial@gmail.com
amigapub@gmail.com
info@amigapress.co.in
Website : www.amigapress.co.in

This edition is published by Amiga Press Inc.

ISBN: 978-93-87295-32-2

First Published 2020

Printed and bound by Amiga Press Inc.

Preface

Plant pathology is the scientific study of plant diseases caused by pathogens and environmental conditions. Organisms that cause infectious disease include fungi, oomycetes, bacteria, viruses, viroids, virus-like organisms, phytoplasmas, protozoa, nematodes, and parasitic plants. Not included are ectoparasites like insects, mites, vertebrate, or other pests that affect plant health by consumption of plant tissues. Plant pathology also involves the study of pathogen identification, disease etiology, disease cycles, economic impact, plant disease epidemiology, plant disease resistance, how plant diseases affect humans and animals, pathosystem genetics, and management of plant diseases. Control of plant diseases is crucial to the reliable production of food, and it provides significant reductions in agricultural use of land, water, fuel and other inputs. Plants in both natural and cultivated populations carry inherent disease resistance, but there are numerous examples of devastating plant disease impacts, as well as recurrent severe plant diseases. However, disease control is reasonably successful for most crops. Disease control is achieved by use of plants that have been bred for good resistance to many diseases, and by plant cultivation approaches such as crop rotation, use of pathogen-free seed, appropriate planting date and plant density, control of field moisture, and pesticide use. Across large regions and many crop species, it is estimated that diseases typically reduce plant yields by 10% every year in more developed settings, but yield loss to diseases often exceeds 20% in less developed settings. Continuing advances in the science of plant pathology are needed to improve disease control, and to keep up with changes in disease pressure caused by the ongoing evolution and movement of plant pathogens and by changes in agricultural practices.

Plant biotechnology is a set of techniques used to adapt plants for specific needs or opportunities. Situations that combine multiple needs and opportunities are common. For example, a single crop may be required to provide sustainable food and healthful nutrition, protection of the environment, and opportunities for jobs and income. Finding or developing suitable plants is typically a highly complex challenge. Plant biotechnologies that assist in meeting the challenge include genomics, molecular-assisted selection, and transgenic crops. These biotechnologies allow researchers to detect and map genes, discover their functions, select for specific genes in genetic resources and breeding, and transfer genes for specific traits into plants where they are needed. NIFA funds research, training, and extension for developing and using biotechnologies for food and agriculture.

—*Author*

Contents

1.

Sustainable Production of Wood and Non-wood Forest Products

INTRODUCTION

Utilization of native woody vegetation is important to the livelihood of the people in arid zones. Foodstuffs, tannins and gum, essential oils, pharmaceutical products are only a few of the numerous non-wood products obtained from woody vegetation. Although called minor forest products, they are of vital importance to the people and often constitute an important part of the total forest revenue.

Of the 350,000 plant species that have been described by botanists, only 3,000 are reported to be sources of useful material for people. Less than 100 of these plants are cultivated on a large scale and none are xerophytic. However, the search for native xerophytic plants of economic value has greatly intensified in recent years.

According to the type of their utilization, non-wood products from dry land vegetation can be divided into the following groups: foliage and fruits, tannins, gums and resins, oils and extracts, fibres and medicinal plants. These products are reviewed in the following sections.

FOLIAGE AND FRUITS

The foliage of woody plants is important to arid zone dwellers: that of palms (Phoenix, Hyphaene, Borassus), for instance, providing raw material for fibres, enclosures, sand-fixing palisades and household articles; that of trees such as Adansonia, Boscia, Cadaba and Balanites providing vitamin- rich food; foliage ofDiospyros melanoxylon, Morus alba and Zizyphus mauritiana, raw material for local industry: cigarette making, sericulture and lac respectively.

The utility of foliage in recycling nutrients, providing shade and reducing wind speed should not be forgotten; well-managed stands of appropriate drought-resistant and palatable deep-rooting woody species also provide valuable aerial fodder for long drought periods when surface vegetation disappears.

Arid and semi-arid zone vegetation comprises a wide range of edible fruit-bearing and food-producing species: Phoenix, Borassus, Hyphaene (fruit, edible pollen and nuts); Grewia, Morus alba, Zizyphus, Tamarindus, Ficus carica, Opuntia, Ceratonia and Olea europea (fruit); Pistachio, Prunus amygdalinus, Pinus, pinea, P. cembroides, P. edulis (nuts). Many of the above play a multiple role in dry zone agroforestry systems, providing soil cover, wind protection, fuelwood and fodder as well as food.

The production and consumption of fruit in arid zones provides a dietary supplement as well as commercial opportunity. The growing of trees for fruit production encourages the preservation of more or less permanent stands or scattered individual trees in otherwise bare lands. Such trees are often a feature of Sahelian landscapes and form the basis of traditional agroforestry land use systems still practiced in parts of that region.

TANNINS

Tannin is produced from the fruit, bark, leaves and roots of many arid- zone shrubs and trees. the preparation of tannin involves breaking or crushing the tannin-rich material which is then washed and boiled with water. after separation of insolubles, the thick, viscous extract is evaporated leaving crude tannin which can be purified by extraction of the crude portion with an alcohol- ether mixture depositing tannic acid. tannins can be of widely varying chemical structure but are capable of converting the gelatin of hides and pelts into insoluble non-putrefying material, that is, leather. tannins are readily soluble in water or alcohol giving strongly astringent solutions also useful in medicines. they are used with feric salts in compounding inks of greenish-black to bluish- black colours.

The practice of tanning hides and skins is extremely important in arid and semi-arid regions where pastoralism is the main land use and wildlife thrives, if protected or adequately managed. Tanning permits the processing and protection of the locally-produced raw materials adding utility and commercial value to a major byproduct of meat production.

A large number of arid and semi-arid zone species yield tannin in commercial quantities. For example, from the bark: Acacia nilotica, A. cyanophylla, Eucalyptus astringent, Parkia biglobosa; from the fruit: Calotropis procera, A. farnesiana; from the wood: A. polyacantha, Schinopsis lorentzii; from roots: Punica granatum,Zizyphus spina-christi.

The styptic and astringent properties of tannic acid are useful in the treatment of inflammations, skin eruptions and bowel conditions and thus form an important principle in the action of many medicinal products.

GUMS

Gums are typical products of broadleaved trees and shrubs. they are complex carbohydrate derivatives of a polysaccharide nature and are either soluble in water as in the case of gum arabic or form mucilages by the absorption of large amounts of water (gum tragacanth). their principal use is in foodstuffs by nature of their ability to impart desired qualities to foods by influencing their viscosity, body and texture; most frequently in confectionery food, flavouring and soft drinks. they also have pharmaceutical applications as demulcents, adhesives in pill manufacture and as emulsifying agents. industrial uses are for adhesives, lithography, paints and inks.

Gums are produced from woody plants either naturally from exudations from cracks in the bark or damage to the bark by insects or animals. Gum flow is also artificially induced by incisions in the bark. The viscous, brittle nodule can be removed by hand.

Gum arabic is the main commercial gum exudate. This gum is mainly obtained from Acacia senegal and some from the related species A. laeta, A. polyaccantha andA. mellifera.

Other gums are gum Karaya from Sterculia urens, S. villosa (India), and S. setigera (Africa); they provide the raw material for emulsifiers, adhesives, fixatives and laxatives. Gum tragacanth from Astragalus spp. of Asia Minor is even more valuable: it is a natural emulsifier in food products such as mayonnaise but is now being replaced, because of its high cost, by synthetic fermentation type products. Gums of commercial interest are also obtainable from the fruit of the carob (Ceratoniasiligua), gum Mesquite (Prosopis latifolia) and Indian Squill from Urginea indica.

RESINS

Natural resins are distinguishable from gums because of their insolubility in water, but because the exudates from so many plants possess this quality, classification of resins is difficult. Resins comprise balsams: resins of a fluid character often used for healing purposes; oleoresins: generally from conifers. These are solutions of resins in essential oils; turpentines also from conifers and some broadleaved species; mastics, such as those from Pistachio spp., used in protecting oil paintings; hard resinssoluble in alcohol and benzene such as “dragon’s blood” and “gambage”; dammars soluble in aliphatic and aromatic hydrocarbons and sandarac, a base for spirit varnishes derived from Callitris and Tetraclinis. Others of the oil-soluble resin group include Copals, oriental lacquers, and substances such as Cashew shell-nut oiland Lac derived from the lac insect. Resins are generally used in adhesives, paper sizing, surfacing, fixtures for perfumes and in medicines. Other resins for religious end uses include: frankicense and myrrh, from the dry-zone species Boswellia spp. and Commiphora spp.

Resin-bearing plants are fairly common in arid zones and the following list provides an indication of their range. USA: Grindelia camporum, Pinus cembroides, oleo-resins for naval stores; Larrea tridentata, resins and turpentines. North Africa: P. brutia and P. halepensis for oleo-resins, Tetraclinis articulata for sandarac.

Latin America: Schinus terebinthifolius and Juniperus californica for medicinal resins and oleo-resins. Sahelian Africa: Commiphora africana and Boswellia dalzielii for myrrh, and frankincense. India: Boswellia serrata and Commiphora urghtii for products similar to those produced by African relatives. Near East: Juniperusmacropodia and Boswellia sacra, oleoresins for perfumery and frankincense.

OILS AND EXTRACTS

Plants that produce essential oils are very characteristic of dry areas, and some have been used since ancient times in the Mediterranean region where the essences of rosemary (Rosemarinus officinalis), thyme (Thymus vulgaris) and lavender (Lavandula angustifolia) are extracted by steam distillation for perfumery.

Eucalyptus oil is steam-distilled from fresh eucalypt foliage obtained from felled trees or cultured coppice shoots. Eucalyptus oils are useful for medicinal purposes (inhalants, embrocations, soaps, gargles, sprays and lozenges), industrial uses (disinfectants, solvents, synthetic thymol and menthol) and perfumery (eudesmil, geranyl acetate, citronella!).

The dry-zone species E. astringens, E. leucoxylon, E. melliodora, E. occidentalis and E. populnea are known to produce oils with commercially high proportions of cineole, geranyl acetate (E. marcarthuri) or citronellal (E. citriodora).

Other oils include those of lauric acid from Salvadora oleoides which provides a substitute for coconut oil; vetiver oil for perfumery from Vetiveria zizanoides; palmarosa oil from Cymbopogon martini var. motia for geraniol; perfumery oils from Rosa demascena and Inula racemosa; zachun oil for soap making from Balanitesaegyptica fruit; karite butter from the fruit of Butyrospermum parkii; jojoba oil from Simmondsia chinensis, a substitute for whale oil; lubricating oil from Jatrophacurcas and-medicinal oils and waxes from Quillaja saponaria, Tabeluia toxofora and Pilocarpus jaborandi. Euphorbia anti-syphilitica is another species producing a white wax, a substitute for beeswax, while the guaynle, Parthenium argentatum produces a latex similar to rubber.

FIBRES

Dry-land woody species offer considerable scope for the extraction and use of fibres for cordage, ropes and handicrafts. The date palm, Phoenix dactylifera, is one such species. It is estimated that about 150 million Phoenix dactylifera can be found in the Near East and North Africa and these also form a potential source of fibre for paper manufacture.

Borassus aethiopium and Hyphaene thebaica provide fibres from pounded leaf petioles from which fibres are manually separated and used for cordage, strings or raw fibres for binding and the manufacture of domestic articles and handicrafts.

In Latin America, Yucca carnerosana, Agave lechuguilla and Agave sisalana are important producers of fibres in dry lands. Agave sisalana, the producer of sisal and a non-wood plant of the Amarillydaceae family, is planted in the dry lands of Brazil at a density of about 5,000 plants per hectare, yielding around 30 leaves per plant per year. The dried sisal fibre, about 6 per cent of the weight of the green leaves, is principally used for the manufacture of baler twine and also rope, cordage and handicrafts (such as macrame). It is also excellent for paper suitable for thin printing papers, cigarette papers, bank notes and teabags. It can be blended with other pulps for newsprint, wrappings and tissue paper.

In India, fibres are produced from the inner bark of the climber Bauhinia vahlii (ropes for domestic purposes); the bark of the shrubby dryland plants Calotropisgigantea, C. procera spp. hamiltonii and Leptadenia pyrotechnica also yield fibres useful for weaving string and nets.

Esparto grass or alpha grass, Lygeum spartum and stipa tenacissima provide important resources of fibre in North Africa yielding 0.2 to 0.7 tonnes per hectare for pulp. They also provide material for handicrafts in the form of baskets, woven mats and screens. Paper made from esparto is smooth with a soft surface possessing excellent printing qualities especially suitable for illustrations and colour with a high dimensional stability.

MEDICINAL PLANTS

Possibly 80 per cent of the world's rural populations are reliant on medicinal plants to maintain their health and to cure their ailments.

Medicinal plants contain a wide range of chemical substances and are very varied in their effects and uses. For instance, diosegenin for wounds and stomach ailments can be obtained from Agave sisalana; steroidal saponins and sapogenins useful as anthelmintics and purgatives from Balanites aegyptica; glycosides and calotropin with strong cardiotonic action from Calotropis procera; the alkaloid artemitin, a potent stimulant, from Artemisia absinthium; astringent, carminative resins fromCommiphora nukul; the car dionotic, antiseptic and gangalgesic stachydrine and other compounds from Capparis decidua; ephedrine, a bronchodilator, from Ephedrasinica; hyoscyamine producing atropine for opthalmology from Duboisia leichardtti; astrogalin, rutin and cardiotonic glycosides from Nerius oleander.

The collection of herbal drugs has long afforded a gainful occupation for many people in the rural areas and the processing of herbal drugs in traditional phytotherapy includes simple operations such as the preparation of powders, pills, lotions, decoctions and liquid extracts.

INVOLVEMENT OF LOCAL PEOPLE

In the arid zones of the world, products derived from the forests provide essential goods and services, such as fuel for cooking and heating, construction material for housing, forage for cattle, and nanwoody products. The forest itself helps to provide clean water and, in many areas, provides shade from a burning sun. In many instances, all these benefits have been taken for granted. However, due to increasing population pressures and lack of appropriate forest management strategies, these benefits are no longer readily available. Increasing pressures on the forests for these and other benefits have resulted in degradation of the vegetative resources and impoverishment of the land. Ultimately, the basic needs of people may no longer be satisfied. The situation can be countered by improved management of natural forests or by establishing forest plantations. But, to do so, people must participate in the design, implementation, and follow-up of these activities. To mobilize this participation, a number of constraining factors have to be understood.

A central aim of forestry programmes is to help people become self-reliant. However, these programmes will not succeed unless they reflect the people's own interpretation of their needs, aspirations, and problems. Therefore, forestry must be forestry for the people and, importantly, involve the people.

CONSTRAINTS AND CONDITIONS FOR PEOPLE'S INVOLVEMENT

Some of the factors to be taken into account in analyzing the place of forestry in a rural economy are summarized in Table. These factors and some possible responses are discussed more fully below.

The Timescale of Forestry

In many cases, attachment to a particular technique of food production is reinforced by considerations derived from the timescale of forestry. Historically, many rural populations have developed a dependence upon outputs of the forest, because the latter existed as an abundant available local natural resource which could be utilized almost at will. As long as the wood products remained abundant, this process of exploitation could take place without regard to the relatively long time involved in producing wood of usable sizes. However, once the point is reached where wood can be supplied only by growing it, the timeframe involved can become an important limiting factor.

Frequently, the timescale of forestry is in conflict with the priorities of the rural people; logically, these priorities are focused on meeting basic present needs. Present needs are imperative, particularly in subsistence situations. Land, labour, and other resources which could be devoted to providing the food, fuel, and income needed today cannot easily be diverted to the production of wood, which will be available only several or many years into the future.

Forestry can continue to exist or be introduced only if it allows the present needs of the people involved to be met. If local tree and shrub cover still exists, it may be possible to provide the same level of production in a less destructive way. In some areas, for example, destructive local cutting of the forest may be halted and reversed by concentrating the cut in designated areas and at designated times, and protecting the rest of the area so that natural regeneration can occur. In other areas, forestry may be introduced in conjunction with other activities which secure income to meet the needs of the farmer until his trees yield products.

The Spatial Distribution of Benefits

Considerations of the spatial distribution of forest benefits are important. To the farmer, the forest is land upon which to cultivate his food and cash crops, a source of fuel and building materials, and a source of fodder and shade. The fact that the trees and shrubs that are utilized could provide the raw material for an economically-important industry elsewhere is of little relevance to him. To expect the farmer to adapt his way of life to accommodate the interests of others is unrealistic. Therefore, the build-up of more stable forest crop/tree systems will only occur when the community, in some way, benefits in appropriate measure from any change in the pattern of natural resources use.

The core of the problem for rural people is usually that they derive insufficient benefits from the forest; in general, this is attributable to conventional forest management objectives and administrative practices, such as an orientation of natural resources towards conservation, wood production, revenue collection, and regulation through punitive legislation and regulation. Consequently, the task of a forester is to engage the people more fully, positively, and beneficially in management, utilization, and protection; this may take the form of greater participation by the people in forest work, development of the income potential of miscellaneous products that can be produced in the forest, or the allocation of forest land for the concurrent production of wood and agricultural crops, or of forestry and livestock.

Cardiotnic, anteseptic and angalgesic stachydrine and other compounds from designed to establish industrial wood crops through farming systems which grow trees or shrubs with food and cash crops. In themselves, the trees and shrubs may bring no direct benefit to the farmer; rather, they are an impediment which complicates his task. These systems will succeed only when the farmer perceives adequate compensation. Land, in itself, is not a sufficient inducement, other than perhaps in the short-term; it has been observed that, over time, multiple-use systems tend to evolve either into settled agriculture (with the rejection of the associated growing of trees or shrubs) or into full- time forestry employment.

Institutional and Technical Constraints

There remain situations in which there is no lack of interest in forestry nor any conflict with other aspects of the way of life, but only a lack of organization or means. Areas which are marginal for agriculture may also be marginal for forestry; this is particularly the case in arid areas, which tend to impose severe climatic constraints on the growing of trees and shrubs. Additionally, arid environments impose other constraints, including that of availability of labour. In arid areas, the planting seasons for both agriculture and forestry are short and coincide. As a result, the availability of labour for tree or shrub planting can be restricted, and planning must allow sufficient flexibility to overcome such a constraint.

Another constraint to forestry in arid environments is the associated cost. Successful tree or shrub planting programmes often involve elaborate site preparation and planting techniques which require sophisticated" and costly equipment. It may be an activity which is beyond the capability and resources of the local community. Therefore forestry, as an activity which the community can implement, is often confined to the manipulation of existing vegetation, with plantation forestry undertaken by the responsible technical branches of government.

Land management on steep upland areas is also likely to present problems of soil stabilization or control of water runoff; establishment of forest cover on parts of the watershed must be accompanied by conservation measures (such as the construction of terraces) to permit stable crop production on other parts. In many instances, farmers will not have the resources to do this. To establish terraces, for example, they would have to forego one crop. Therefore they will need external support, such as credit and food aid.

Another institutional issue is security of land tenure. Patterns and traditions of use on tribal or communal land often make no provision for usages such as forestry, which require the setting aside of land for a particular purpose for relatively long periods of time. Farmers must have assurance that they will control the land on which trees or shrubs are planted.

Frequently, the lack of a tradition in forestry extends beyond just a lack of knowledge about growing trees and shrubs or of an appropriate institutional framework within which to implement it. Usually, it contrasts with a "deep" tradition in agriculture. This contrast is reflected in the attitudes the people have towards forestry, which are significantly different from attitudes towards agricultural crops and livestock. A forest often tends to be seen as a negative element of the environment by farmers. To the settler, it is an impediment to the clearing of his lands and a haven for his enemies. These views can persist long after the forest has receded from the immediate vicinity of a farm or community. Hostility towards forests and trees can persist in areas which already experience shortages of fuelwood and building poles, because of the damage done to agricultural crops by birds which roost in trees.

Other institutional constraints can also limit the success of community forestry approaches: poor seedling distribution systems, poor management, lack of village funds in order to pay for labour, lack of labour at the right time, planting ill-suited exotics vis-a-vis other options for natural stand management, and so on.

Providing protection to communal woodlots is a particularly frequent problem which may sometimes be reduced by staggering tree planting over several years so that the area which must be protected at any time is kept as small as possible. When the trees have reached a stage where they will be safe from livestock damage, the area can be opened for grazing and browsing. People might also be given access to protected areas for fodder collection, and it is quite likely that fodder production will be greater on these lands if they are protected from free grazing.

The type of silvicultural management system which is adopted is also important. For example, if the community's on-going needs are to be met, annual cutting cycles may be required. However, if the woodlot is very small and the size of the community large, it may not be feasible to divide a small harvest on an annual basis. Similarly, if local people are to be involved in the harvesting, socially workable schemes such as clear-cutting may be preferable to more technically difficult systems like selective thinning. Any management system must, of course, be within the technical competence of the user group and be simple enough that members will feel confident that they can control it in practice.

HIGHLIGHTS OF SECTION

With increasing population pressures, the forest products which were taken for granted in the past are no longer as plentiful. With their strong tradition in agriculture, rural people are sometimes reluctant to practice forestry on their lands. However, rural dependence on forest products (such as fuelwood, lumber, food, income and employment) is a reality. For a tree or shrub planting programme to be successful, consideration of the present and future needs of the people, as well as their prejudices and traditions, is necessary. Because there is competition among agriculture, livestock grazing, and forestry developmental activities, the local people must be persuaded that multiple-use programmes (which are realized only after a period of years) have value.

Occasionally, there is an institutional constraint to the implementation of the tree and shrub planting programmes; such examples include situations where there may be lack of organizational structure or security of land tenure. The relatively high costs of some endeavours and the technical skill needed also can be prohibitive. Lack of understanding of the role of trees and shrubs in enriching the environment or a reluctance to change can hinder attempts to combat desertification with forestry.

❐

2.

The Plant Parasitic Green Algae

Among the autotrophic plants at least — and possibly within the whole of the plant kingdom, the algae are unexcelled for variation in structure, range of habitat and diversity of role. There are excellent texts dealing with their structure and reproduction and much has been written about their ecology; but it is not easy to locate assembled information on their roles in the biological world and of their impact on man, plants and other animals.

They affect us in numerous and unsuspected ways, and only after collecting much information on this diversity of role will we be able to construct a full panorama of just how important they are. Let us not underestimate their importance on the grounds that most are microscopic. What they lack in size, they make up for in numbers — and herein lies the secret of their success in performing tasks which appear to us incredible.

It would be much easier to comprehend this diversity of role-playing among algae if we could classify these roles in such a way that groupings by function or habitat corresponded with physiological differences. Unfortunately this classification cannot yet be made. Our ignorance of algal physiology at the present time prevents our performing this synthesis of natural fact and fundamental function.

However, an attempt should be made to correlate these reported natural facts if we wish to obtain an overall picture of the significance of algae. Appreciation might also be quicker and more embracing were we to devise a framework on which to hang the various fragments of this bolus of factual matter. In the absence of a natural system of classification, we must construct an artificial one; and perhaps the simplest springs from man's anthropocentric attitude — since he fancies himself the centrepiece of the biological world and all things else as useful to him or not.

Adopting the criteria of utility and nuisance, then, we have a temporary framework for assembling information on the biological and economic possible; and at least this is a start in marshalling into some order the widely scattered pieces of a mosaic never yet assembled in its entirety. Although this framework is highly artificial and will no doubt be imperfect, it can serve meantime as an assembly line for attaching bits and pieces of information as they come to hand from the production lines of literature and experience. When sufficient data have been collected, it may be possible to reclassify these roles using criteria more closely related to fundamental functioning. This is to be hoped for and will be very welcome if achieved because then we may be able to discern some connections between cause and effect. In the meantime we must press on with the proposed system in all its artificiality.

Rarely does a system of biological classification work perfectly — generally some compromise must be made. It would be unwise therefore to use this framework as a Procrustean bed to which all relevant detail must be racked and made to fit. This could result in distortion. It is better to include one more category to accommodate those items which do not readily fall within the compass of the existing headings. These misfits are to be found appropriately enough under the title of 'Flotsam and Jetsam.'

So, the major breakdown of the subject matter appears as follows — with some indications of sub-headings:—

- Algae in useful roles.

  e.g. as a source of organic matter, food, chemicals; in soil colonization; etc.

- Algae in nuisance roles.

  e.g. in water contamination; as a cause of poisoning; as parasites; etc.

- Flotsam and Jetsam.

  e.g. thermal algae; algae in art and medicine; etc.

Finally, the reader must always bear in mind that this series of articles could not be written were it not for the fact that many other scientists have done the basic research work and published their results. Theirs has been the more arduous job. Perhaps the greater magnitude of their part as distinct from the writer's can be set in its true perspective in the manner of Montaigne who once wrote: 'I have gathered a posie of other men's flowers and nothing but the thread that binds them is my own.'

ALGAE IN USEFUL ROLES

Algae as Producers of Organic Matter

Since algae are autotrophic plants, they share with the remaining more advanced autotrophic members of the Plant Kingdom the capacity to synthesize complex organic molecules out of carbon dioxide and water, using biochemical kingpin of life on our planet — is called photosynthesis. On the unceasing performance of this process hangs the continuity of all animal life— which is utterly dependent on the plant world. Directly or ultimately, land animals depend on land plants for food and aquatic animals rely similarly on algae as their basic source of chemically-bound sunshine and hence energy.

Let us diverge for a while to take a look at the very fundamental implications of this process called photosynthesis — the prerogative solely of the plant kingdom. This is the term applied to a set of reactions carried out in green plants whereby a simple gas of the air, carbon dioxide, is converted into much more complex compounds among which sugars, starch and oils are most important. Carbon as it exists in carbon dioxide is in its most highly oxidised state whereas it is partially reduced when found in sugars, starch, oils and indeed every form of organic matter. To synthesize these reduced complex compounds of carbon the plants require energy, and this comes from sunlight.

Animals including ourselves and the non-green plants such as fungi consume this organic matter in some form or other and in the process of digestion break these compounds down again. The purpose of this break- down is to release the built-in sunlight energy which is then used to drive the cellular machinery and life processes of the animal kingdom and the non- green plants. So, plants photosynthesize organic matter and in doing so harness energy: animals degrade (or 'desynthesize') organic matter and release energy.

We must regard organic matter not only as a hotch-potch of complex organic chemicals but as a colossal lay-by or stockpile of available energy for the biological world. The more that is stock-piled the greater the animal populations that can be sustained, so that as far as humans are concerned the interdependence of this energy stock-pile and animal population is synonymous with food supply.

The existence of animal life on earth is related to the amount of reduced carbon. In terms of a bank-balance of reserve biological energy, the amount of credit is synonymous with the surplus of reduced carbon over and above that used by animals and non-green plants. One cannot envisage a time or circumstance when the world will be 'in the red' for reduced carbon; but on one point we can be quite sure — we will never be alive to see this happen because all life on earth will quickly disappear the moment an overdraft of this commodity is established.

From this detour in our story, a couple of points emerge which must be kept in mind so as to get the correct perspective on what is to follow. Firstly, green plants (or more correctly, those containing the pigments capable of initiating photosynthesis) are the only organisms in the world which can synthesize complex organic matter from simple inorganic substances in sufficient quantities to bring about a surplus: and secondly, surplus of this organic matter is synonymous with available energy on which the rest of the biological world is absolutely dependent. It is necessary to reorient our thinking along these lines; and having done this, we are in a better position to appreciate the fundamental importance of the algae in the biological scheme of things.

The proportion of our earth occupied by permanent water — fresh and marine — exceeds 71 per cent of the total earth's surface; and except for a few freshwater ferns and seed plants, this is the undisputed and botanically uncontested domain of the algae, since they are predominantly water-plants. Because the algae are photosynthetic they must be synthesizing throughout this aquatic environment enormous quantities of organic matter — in other words, their contribution to the credit side of the biological bank-balance of energy must be colossal.

This association of the ideas of the aquatic habit of algae and the vast aquatic surfaces of the earth leads to the posing of two questions: the first is —

'How much organic matter is produced by the algae?', and the second — 'What happens to all this organic matter?'. The answer to the first question forms the subject matter of this present topic; but the answer to the second is much more diffuse and will be found scattered throughout the rest of this series of articles. This first part is designed to enable you to achieve some realization of the quantities of organic matter estimated to be produced by algae, and meanwhile only sufficient reference will be made to its fate to fill out the story and set the scene for what comes later.

This organic matter can be divided into two major groups — as follows:—

- Materials which are retained by the algal cells and built into cell material and food reserves for increase in cell size and reproduction
- Materials which are not retained by the algal cells but are excreted sometimes to the benefit or even detriment of other organisms.

Most would correctly guess that of the products of photosynthesis the bulk is used by algae to build cellular material and food reserves; but it may cause surprise to be acquainted with the fact that many algae excrete organic matter. This is not unexpected really since the medium which supports them and supplies their nutrients must also act as their toilet for unwanted metabolic end-products. Many algae produce large quantities of mucilage and slime on the outer surfaces of their cells, and as this is being removed from the outside by the ambient water it is continuously being replaced from within. These mucilages are carbohydrate in structure. But certain algae excrete small quantities of substances other than carbohydrates; and while we do not know the identity of many of these chemicals, at least we are aware of their presence because of the effects they produce. Several examples of excretion will be briefly cited now for the purpose of pencilling in a few of the broad outlines of what to some will form a new and large panorama of unknown scope:—

- The product of ectocrine substanceswhich affect the matabolism of other algae and animals
- The excretion into freshwater of chemicals referred to as 'tainting chemicals'. These diffuse out of the algal cells into drinking-water in holding areas and produce those smells and off-flavours so well known to water-works engineers
- The excretion of substances which act in an algistatic or antibiotic fashion
- The production of substances excreted into stream waters which enable migratory fish to 'home' on the stream and scene of their nativity
- The formation of chemicals highly toxic to fish, birds and other vertebrate animals
- The excretion of synthesized nitrogenous compounds by nitrogen- fixing blue-green algae.

All this material, cellular and excreted, is adding metabolic currency to our planet. Most of it supports active life and living organisms, and the excess — if it serves no other immediate purpose — contributes to the savings of reduced carbon in the world's bank balance of reserve biological energy.

It is not intended to expand our treatment of any of the above effects at the present moment because this will be done later according to where each effect comes in the overall plan set out earlier; the intention is to spend some time discussing how much cell material and organic matter is produced by algae, because this will enable us to acquire some comprehension of the importance of this group of plants in the most fundamental aspect of plant life. Information will also be included on the chemical nature of compounds known to be excreted.

Much effort has been expended in an endeavour to estimate the total amount of organic matter produced by algae in the sea. About 71 per cent of the earth's surface is occupied by saline seas, and therefore we can say that about 71 per cent of the earth's surface is inhabited by a marine algal flora. Higher plants are restricted to the remaining 29 per cent wherein they mostly occur on land or extend into very shallow brackish water. Returning to full- strength sea-water, we must realise that this is inhabited by algae as sole representatives of the photosynthetic section of the plant kingdom. The algae may be attached as seaweeds or free-floating as plankton; but according to Ryther (1960) 99 per cent or over of the oceans are too deep to allow attached algae to grow and although some of the greatest concentrations of vegetation known are to be found in littoral algal stands, the littoral algae pale beside the planktonic as producers of organic matter.

To measure the total organic matter manufactured by planktonic algae is a difficult task and several attempts have been made using a number of different methods. As a group, the algae possess several different forms of But one chlorophyll of the pair is common to all divisions, namely, chlorophyll 'a'. So methods of determining the concentration of phytoplankton havecentred on estimating the amount of this particular form of chlorophyll.

During the International Geophysical Year many determinations of chlorophyll 'a' were made as part of the oceanographic programme. Values for the upper illuminated layers of the Atlantic Ocean were found to fall within the range of 0.1-0.5 mgs./cubic metre of sea-water — with an average value of maybe 0.25 mgs. Ryther then calculates as follows:— assuming the ratio of chlorophyll 'a' to dry weight (D.W.) of phytoplankton to be 1:100, 4" the D.W. of plankton/sq. metre to a depth of 100 metres (a 'round' figure for the depth of ocean to which light will penetrate) would be $0.25 \times 100 \times 100/1000$ gms. =

2.5 gms. on the average. He adds a 'generous 0.5 gms. to allow for the richer coastal waters and to include the benthic algae.'

This added to the 2.5 gms. for plankton gives us a figure of 3.0 gms.

4" the oceans support an average standing crop of about 3.0 gms. D.W. of algal material/sq. metre of surface to a depth of 100 metres.

The saline areas of the world total 361×106 sq. kilometres of surface.

4" the total figures for the D.W. of algal cellular material for the saline seas of the world are:

$(361 \times 10^6) \times (3.0 \times 10^6)$ gms./sq. kilometre. = 1.1×10^{15} gms. = 1.1×10^{12} kgs. = $(1.1 \times 10^{12}) \times 2.2/2240$ tons = $(1.0^7 \times 10^9)$ tons D.W.

Although we might want to class this figure as a 'guestimate' (because of the large approximations made) it does represent a working assessment which gives some idea of the algal biomass of the world's oceans. This is often called the 'standing crop'. The term is applied because the algal biomass is in essence a crop replete with food reserves of starch and oil. It is grazed by zooplankton and therefore acts as a starting point for many food chains.

It is interesting to set this table alongside table for land production. Ryther has also assembled this information. Various people have determined the relative proportions of the earth's land surface according to plant cover — these proportions are as follows:—

Wasteland (desert, arctic regions, mountains) 50 per cent, Cultivated land, grasses, sedges, brush, etc. 20 per cent Forests 30 per cent.

We can neglect the first category because of its paucity of vegetation: so 50 per cent of dry land contributes virtually nothing each year in terms of plant production. The contribution of the second category is worked out as 1.5×10^{13} kgs.; and that of the third category as 1.1×10^{15} kgs. We can now get some idea of the standing crop of plant material:—

Oceans		1.1×10^{12} kgs.
Land —		
	Wasteland	0
	Crops, grasses	1.5×10^{13} kgs.
	Forests	1.1×10^{15} kgs.

Ryther then points out that the productive land plants, occupying one- eighth of the area inhabited by marine algae, maintain a biomass more than 1000 times greater.

'What of the productive capacity of the land and sea? Does it follow that the algae are equally insignificant in the annual production of the earth's organic matter?' These figures have been compiled also by Ryther and their summary is as follows:—

Ocean		3.6×10^{13} kgs./year.
Land —		
	Wasteland	0
	Crops, grasses	3.0×10^{13} kgs./year.
	Forests	2.2×10^{13} kgs./year.

These table represent net production i.e. the excess of a plant's production over its espiratory and reproductive requirements.

'The interesting fact which emerges from all this is that the annual rate of organic production on land and in the sea is about the same despite the fact that the latter is accomplished by a flora less than one thousandth the biomass of the terrestrial vegetation. The explanation for this is that most of the bulk of land plants is in the form of slowly-growing, non-photosynthetic structural tissue' (Ryther 1960). This gives us some idea of the marine algal contribution in terms of material retained within the cell.

Now let us consider the material excreted by the cell. If we take a sample of fresh or sea-water, it is possible to show by chemical means that it contains organic matter in true solution. This organic matter can be produced either by decomposition of plant and animal bodies or by direct excretion by living algae and animals as a normal function of their metabolism. It has been quoted that a figure for the total amount of organic matter estimated in the oceans and it is of the order of 6.75×1012 tons i.e. 6,750,000,000,000 tons. No insignificant amount of soluble marine compost!

Commenting on the loss of organic matter from algal cells, the same author wrote — 'It seems unlikely that the organization of any cell can be so perfect as to prevent entirely the escape of the soluble organic substances which it contains to the surrounding water, and evidence is accumulating which shows that algal cells do, in fact, 'leak' to an appreciable extent'. It is not difficult to give examples of algae which excrete appreciable amounts of organic matter.

Chlorophyll—a pair of Chlorophyll being found in each of the large divisions on the surface of fish-ponds — notable for the amount of slime on the outside of its filaments. In fact, the presence of slime is a rough and ready diagnostic feature of this particular alga. Among the brown seaweeds Macrocystis should be mentioned as a slime-producer, along with many of the red seaweeds. Some blue-greens are continuously shedding mucilage from the outside of the cell wall. In diatoms with a raphe, the cytoplasm is in contact with the ambient medium — there being only a thin membrane separating the cytoplasm from the aqueous environment.

When estimates of standing crop are made by determining chlorophyll 'a', the figure gives an index of the accumulation of cells only — the living biomass. We have no way of knowing how much organic matter has leaked out up to the time of sampling and will leak out during the life-time of the algal plankton. Put in another way, the 'standing crop' figure gives us a rough idea of intracellular organic matter but not the faintest clue to the amount of extracellular organic matter that has been lost up till and after the time of sampling.

For fisheries purposes, the intracellular figure is the major one of significance; but for our purposes trying as we are to formulate some idea of the algae as producers of total organic matter, the extracellular figure would be necessary to know.

As yet there seems no way of estimating this. All we can say is that some fraction of this 6.75 × 1012 tons of organic matter in the sea represents leakage and excretion from algal planktonic cells.

What is the chemical identity of these excreted compounds? Over the last decade many algae have been studied in the laboratory in pure culture; and with our improved analytical methods, we have been able to identify some of their metabolic end-products much more accurately than before. A lot of work has been done in this field and Fogg (1962) in summarizing this research has divided the extracellular products in the following way depending on their chemical nature:

- Acids:

  lactic and succinic, under semi-aerobic conditions; by Prototheca zopfii.

  acetic and lactic — anaerobic conditions; by Chlorella.

  oxalic, tartaric, succinic; by Oscillatoria.

  glycollic, oxalic and pyruvic; by Chlamydomonas.

One interesting reference is to the work of Tolbert and Zill (1956) who cultured Chlorella pyrenoidosa with a radioactive source of carbon and found that with only one minute of illumination this alga loses to the medium between 3 and 10 per cent of the carbon it fixes. This 'lost' carbon appears in the form of glycollic acid.

Long-chain fatty acids are also produced.

- Amino Acids and Peptides:

Certain blue-green algae liberate considerable amounts of nitrogenous extracellular products in the form of peptides and compounds containing amide-nitrogen: but free amino acids are generally found in small quantities only. One writer reported the presence of alanine, glutamic and aspartic acids in the culture fluid of Calothrix brevissima (Watanabe). Fogg (1959) quoted the amount of nitrogenous product excreted by Anabaena as 10 per cent of the increase in normal nitrogen of the cell contents — and 20 per cent or more

when the culture fluid was deficient in iron. The phenomenon of nitrogen excretion is not confined to the nitrogen-fixing blue-greens, because the same sort of thing occurs with non-nitrogen-fixing members of this group. Nor is the phenomenon confined to the blue-green algae. It has been reported for Chlorella pyrenoidosa, Chlamydomonas spp, Tribonema aequale, Navicula pelliculosa and even in the brown alga Ectocarpus confervoides.

- Carbohydrates and Mucilages:

Oscillatoria splendida excretes polysaccharides into the medium; and Anabaena cylindrica loses to its growing medium a complex polysaccharide which consists of glucose, xylose, glucuronic acid, galactose, rhamnose and arabinose in the molar ratios of 5:4:4:1:1:1 . In the genus Chlamydomonas, 18 species were found by Lewin to produce some sort of polysaccharide — which in most cases consisted mainly of galactose and arabinose. As mentioned earlier, Spirogyra, Macrocystis and many algae excrete mucilage continuously.

• Vitamins and Growth-Influencing Substances:

Thiamin (vitamin B1) has been detected in the culture fluid of a species of Coccomyxa by Lewin. A number of algae including Chlorella, Anabaena cylindrica and Oscillatoria apparently excrete chemicals of the auxin-type of plant hormone. Many algae produce chemicals which influence the growth of others — either through stimulation or inhibition (Hartman). As far as inhibition is concerned, this may even be auto-inhibition — comparable possibly with 'staling' in fungal cultures. This inhibitory action can take several forms and Hartman has listed a number of observed effects caused by algal metabolites — such as

- Interference with the dark reaction in photosynthesis
- Reduction of oxygen consumption of bacteria
- Blockage of mitosis.

  To these can also be added the adverse effects on animals of algal extracellular metabolites — such as

- Interference with the filtering action of Daphnia (Hartman)
- The poisoning of fish and other animals — even possibly humans
- The effect on the filter-feeding rate of oysters of 'rhamnoside' — a suspected extracellular metabolite of algae.

Referring more specifically to antibiotic substances in the popular meaning of this word, we find reports of Chlorella pyrenoidosa filtrates exhibiting antibacterial activity; and this has been shown due to oxidation products of unsaturated fatty acids. Oscillatoria splendida, Stichococcus bacillaris and Protosiphon botryoides are also reputed to produce antibacterial substances. Working with Phaeocystis pouchetii a common marine planktonic unicell, Sieburth put forward evidence that the thermolabile antibiotic which he found in the culture fluid was acrylic acid.

(v) Enzymes: Harder in 1917 showed that a species of Nostoc was able to use starch and other high molecular weight compounds as a source of carbon. This implies that an extracellular enzyme must be excreted to hydrolyse these sources of carbon down to smaller assimilable substances. Similarly, Nitzschia putrida is thought to do the same thing because Pringsheim showed that it is able to liquefy gelatine: in this case the enzyme must be proteolytic.

All this evidence demonstrates the excretion of chemicals under laboratory conditions; but does excretion occur in nature? Fogg conducted some experiments in Lake Windermere to determine the amount of carbon excreted by algae under natural conditions. He filtered samples of lake-water to which C14-bicarbonate had been added and determined the amount of C14-labelled organic matter in the filtrates.

The experiments, carried out at times when diatoms were the most abundant component of the algal plankton, showed the apparent liberation as extracellular products of 3 to 90 per cent of the total carbon fixed. The proportion of extracellular product tended to increase with depth, and the main form of excreted product seemed to be glycollate. Glycollic and other acids have been detected in sea-water by Koyama and Thompson; and Fogg and Nalewajko have estimated glycollic acid in concentrations of the order of 1 mg./ litre (1 p.p.m.) in both sea-water and freshwater.

We also know that other substances as yet unidentified are produced by many freshwater algae. The water-works engineer is only too familiar with taints and off-flavours of water supplies. These effects can be produced by Synura, Dinobryon, Aphanizomenon, Fragillaria, Asterionella, Peridinium — to mention but a few.

Virulent toxins which cause the death of higher animals such as fish, birds and cattle, are also produced by some algae in standing or stagnant freshwater. While the nature of these compounds is not known, as least by their effects we know they are produced. In some cases laboratory work has definitely shown that these toxins are excreted — as with Prymnesium parvum, Microcystis and Anabaena (Prescott): but for others there is the possibility that the toxin may be released only after cell disintegration.

Despite the examples quoted, it seems reasonable to say that of the 6.75 × 1012 tons of organic matter calculated to be in the seas the major part would be contributed through the decay of plant and animal body; but it must be remembered that an unknown proportion of this total is due to compounds excreted by the planktonic algae. Some of the more noticeable examples of effects of excreted metabolic products occur in freshwater and small bodies of standing water — and not in sea-water, with the exception of products released by red-tide organisms. So while the effects of these toxins and other substances may be startling, the amount of organic matter involved need not be very great.

Algae, planktonic and attached, occur in freshwater also; and in lakes, ponds, rivers and rills make their ceaseless contribution to the world's organic matter. But again it is impossible to assess the extent of this contribution. The sea is a reasonably uniform medium as a chemical environment (apart from the seasonal fluctuation of nitrogen, phosphorus, silicon and one or two trace elements) and extrapolation of analytical figures can be made on a world- wide basis. But the contribution of freshwaters cannot be assessed in quite the same way. There are numerous barriers to this.

Due to suspended colloidal matter, the depth to which light penetrates in many if not most freshwater areas would not be anywhere near the 100 metre figure of the open ocean areas. In fact it has been stated that light will penetrate only to about 10 to 15 metres in freshwater. Many areas in any case would not attain 100 metres in depth. So when we realise that the total area of freshwater lakes and rivers is roughly 0.7 per cent of the total area of oceans and when we take into account this low light penetration in freshwater, it would appear that they are no match for the oceans in the production of cellular material. Yet it must not be forgotten that many freshwater unicellular algae including diatoms are very much larger in cell size than those of oceanic plankton and there are many multicellular algae in freshwater plankton which do not exist in the seas. Desmids for instance, a common member of freshwater plankton, do not occur in the seas at all.

The composition of freshwater varies enormously and differs with location and local climate, surrounding geological formations and types of land cover. Some of these bodies of water are 'hard' whilst others are 'soft' — with correspondingly different algal floras. Desmids for instance in England are associated with a high ratio of sodium + potassium to calcium + magnesium.

Griffiths wrote that 'water blooms are more common in large bodies of water especially in those on the newer geological formations; for this reason they are uncommon in the larger British lakes, most of which are situated on old geological areas. Water-blooming is very common on the rest of the great North European Plain of the Continent.' The pH of water is another variable to be reckoned with. Again, rivers have an algal flora all the year round: due to constant flow, there can be no thermal stratification to act as a regulator of growth as found in large lakes and oceans.

Another problem associated with the fresh water habbits is the assessment of the production of extracellular organic matter. We would have no way of determining the amount of this commodity since now the organic matter figure could be grossly affected by what leaches from the land plus the debris and humus which comes from terrestial plant life on the banks of these fluviatile areas.

So, in the face of all these 'if's, of's, and's and but's' and conflicting variables, it appears we have to abandon all hope of arriving at figures for intra- and extracellular organic matter production of freshwater algae. Before closing however, we should mention that algae grow and proliferate on ice and in snow, and areas such as the Arctic and Antarctic despite their cold and inhospitable habitat are not without their algal contribution to the world's organic matter.

Neither must we forget the world's mantle of soil since this supports an algal flora also. Any attempt at assessing the contribution of organic matter from algae in this medium is at present impossible — and may remain that way. Certain types of cultivated land such as paddy fields would have a high figure for both forms of algal organic matter; but many others such as the dry deserts would be wanting in it. However, in between these extremes there is a lot of soil harbouring an algal flora that is imperceptibly building up the level of carbon compounds.

Perhaps it is time now to attempt distillation of a few pure ideas from this crude liquor of facts and figures. The total area of the seas is much greater than the total area of freshwaters and it is unlikely that the algae of freshwaters produce as much organic matter in toto as those of marine habitats; but these freshwater ones do produce some — how much, is anybody's guess. As in the seas, there are extensive food-chains in freshwater environments which have their starting point with planktonic algae. So, alongside Ryther's figure for cellular organic matter production by marine algae, we are unable to put a similar figure either for freshwater or for soil algae; and we have no idea of the amounts of excreted organic matter of either freshwater, marine or soil algae.

In tabular form the situation may be summed up as follows:— Intracellular organic matter — marine = 1.07×10^9 tons D.W.

" " " — freshwater = ? " " " — soil = ?

Extracellular organic matter — marine = ? " " " — freshwater = ?

" " " — soil = ?

Total algal contribution to the world's organic matter = 1.07×10^9 tons D.W. +

And this is where we have to leave the organic matter story at the present time — one 'guestimate' and a gaggle of question marks. Consequently we cannot even hazard a guess for the total algal contribution to the world's organic matter production. But having in mind the dictum of The Ancient Mariner — 'Water, water, every where, ...' and remembering that the presence of water in its varying forms is almost synonymous with the presence of algae, we can say without question that the algae would undoubtedly be the greatest contributors to the world's lay-by of available biological energy in the form of organic matter.

However, there is one further point to bring out before closing. On land, our organic matter production is subject to great inroads by parasites — we merely harvest what the birds, the rodents, the insects, the nematodes, the fungi, the bacteria, the viruses, the weeds and the climate leave us. Insects alone are said by some to ravage a third of our crops.

On land, nothing living (except bacteria) seems free from insect attack — they have invaded every ecological niche and preyed on every form of life. The only niche that has been hostile enough to prevent colonization is the sea. While the marine algae have not escaped the ravages of bacteria and fungi (and possibly viruses), at least they are not devastated by insects. Presumably therefore, the wastage of the sea's annual organic matter production in terms of cellular material could be extremely low; but unfortunately this cannot be said for the land's production — a major point in favour of the algae.

GREEN ALGAE

Cephaleuros species are filamentous green algae and parasites of higher plants. In Hawai'i, at least two of horticultural importance are known: Cephaleuros virescens and Cephaleuros parasiticus. Typically harmless, generally causing minor diseases characterized by negligible leaf spots, on certain crops in moist environments these algal diseases can cause economic injury to plant leaves, fruits, and stems. C. virescens is the most frequently reported algal pathogen of higher plants worldwide and has the broadest host range among Cephaleuros species.

Frequent rains and warm weather are favourable conditions for these pathogens. For hosts, poor plant nutrition, poor soil drainage, and stagnant air are predisposing factors to infection by the algae.

Symptoms and crop damage can vary greatly depending on the combination of Cephaleuros species, hosts and environments. The most severe disease combination in Hawai'i is C. parasiticus on guava (Psidium guajava).

Spots caused by C. parasiticus on guava leaves are a top-down, intercellular, full-thickness necrosis, destroying both upper and lower epidermal cell layers and all intervening tissues. Then, a bright but relatively sparse orange algal bloom bursts forth from the undersides of with C. virescens, the more commonly occurring species in the genus.

The distinctive and much more common spots caused by C. virescens on many hosts in Hawai'i appear on the upper leaf surface as raised yet flattened, textured, burnt-orange to brown or rust-coloured, circular areas up to about 2 cm in diameter, having indistinct, filamentous margins and fuzzy surface topographies. Yet, this species is generally a relatively harmless, subcuticular leaf parasite.

THE PATHOGEN

The disease is called algal leaf spot, algal fruit spot, and green scurf; Cephaleuros infections on tea and coffee plants have been called "red rust." These are aerophilic, filamentous green algae. Although aerophilic and terrestrial, they require a film of water to complete their life cycles. The genus Cephaleuros is a member of the Trentepohliales and a unique order, Chlorophyta, which contains the photosynthetic organisms known as green algae.

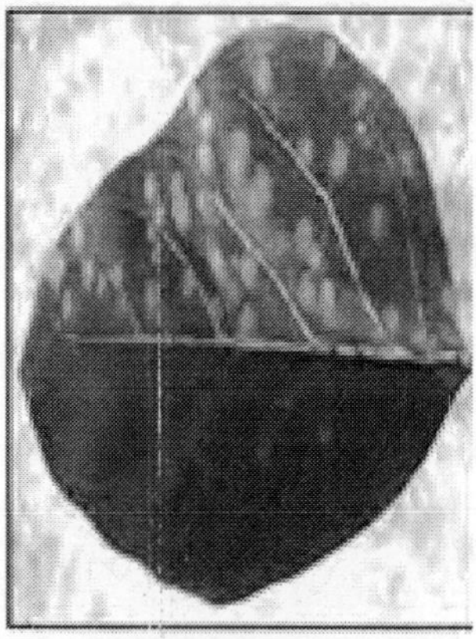

Fig. Algal leaf spot of avocado (Persea americana), caused by Cephaleuros virescens Cephaleuros species consist of branched filaments that comprise a thallus in the form of irregular discs.

The thallus grows below the cuticle or sometimes below the epidermis of the host plant. This pigmented thallus (orange to red-brown) consists of a prostrate portion that is branched irregularly with irregular cells and an erect portion of unbranched hairs, with cylindrical cells, sterile or fertile, protruding through the cuticle. Haustorial cells are sometimes present inside the plant host's tissue. Sporangiophores bear one or more head cells subtending sporangiate-laterals. Gametangia are terminal or intercalary on the prostrate cell filaments.

Cephaleuros species are found in tropical and subtropical climates, on all continents and probably all islands between about 33°N and 33°S of the equator, provided the temperature and humidity are suitable for their growth and reproduction. On the U.S. Gulf Coast, C. virescens has been recorded on at least 287 plant species and cultivars, including 80 that are subject to stem infections. Cephlaeuros species are capable of both asexual and sexual reproduction. Through sexual reproduction, the resulting zygote produces a dwarf sporophyte. The life history consists of alternation of heteromorphic generations, with the sporophyte reduced to a dwarf plant.

The asexual stage is probably much more important to the typical infection and disease processes. These pathogens are dispersed by wind and splashing water. On leaves, symptoms vary according to the Cephaleuros-host species combination. For C. virescens, leaf spots are usually on the upper leaf surface, raised, velvety in appearance, in shades of orange or brown. Leaf spots caused by C. parasiticus on guava are topdown, full-thickness necrosis, destroying both epidermal layers (abaxial and adaxial) and all intervening tissues.

The bright but relatively sparse but brilliantly pigmented orange algal bloom occurs on the undersides of leaves, rather than the upper. On some guava types a chlorotic, yellow halo surrounds the lesions. On guava fruits, spots are slightly sunken and cracked and when severe lend a scabby appearance to the fruit skin. On twigs and branches, lesions may lead to dieback and loss of distal organs.

CEPHALEUROS DISEASE CYCLE

Hosts are inoculated when sporangia or thallus fragments with sporangia are deposited on susceptible host tissues. Infection occurs and symptoms develop under moist conditions when motile zoospoores are released from the sporangia, penetrate the host cuticle, and generate disc-like algal thalli with threadlike algal filaments. The effects of the spots on plants are reduced photosynthetic area of leaves, defoliation, loss of fruit marketability, twig dieback, and tissue necrosis. The pathogens reproduce and survive in spots on leaves or stems and in fallen plant host debris.

Fig. Algal Leaf and Fruit spot of Guava caused by Cephaleuros Parasiticus

HOSTS

Species of Cephaleuros are very common on the leaves of such economically important tropical trees and shrubs as tea (Camellia sinensis), kava (Piper methysticum), pepper (Piper nigrum), magnolia (Magnolia grandiflora), coffee (Coffea arabica), oil palm (Elaeis guineensis), avocado (Persea americana), vanilla (Vanilla planifolia), mango (Mangifera indica), breadfruit (Artocarpus altilis), guava, coconut (Cocos nucifera), cacao (Theobroma cacao), as well some citrus (Citrus spp.) cultivars. Cephaleuros species do not affect certain key subsistence crops in the Pacific, such as banana (Musa spp.) and taro (Colocasia esculenta), although coconut and breadfruit are hosts for leaf spots.

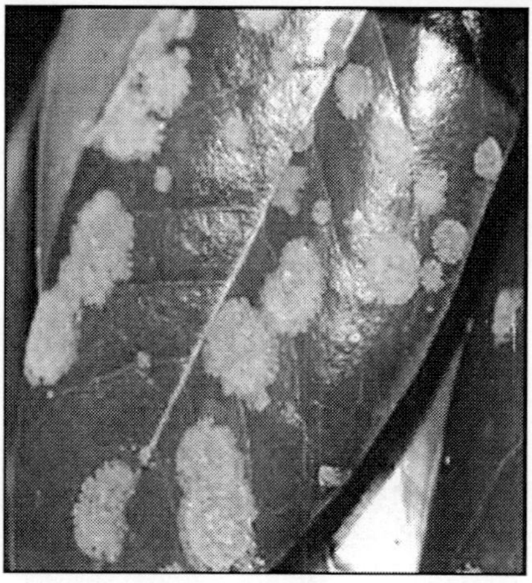

Fig. Algal Leaf Spot of Magnolia (Magnolia Grandiflora) Caused by Cephaleuros Virescens

CROP DAMAGE

Generally, on most hosts, leaf infection is of little direct economic importance and is confined to low-hanging branches near ground level. On more susceptible hosts such as guava there can be significant tissue necrosis, extensive defoliation, economic injury to fruit quality,crop loss, and reduced photosynthesis and plant vigour.

We may separate Cephaleuros species and the damage they cause into two groups on the basis of their mode of parasitism: subcuticular vs. intercellular. C. virescens is a subcuticular parasite, while C. parasiticus is an intercellular parasite. This latter mode of parasitism permits more tissue damage than subcuticular parasitism.

In Hawai'i, Cephaleuros damage to limbs, twigs, and shoots does not seem to be important as in some other locations. On carambola (starfruit, Averroha carambola) in Florida, twigs and limbs may be attacked by C. virescens. Symptoms include rough circular greenish-grey or rusty-red areas and shoot dieback. Most fruit damage from Cephaleuros in Hawai'I probably occurs on guava and is associated with C. parasiticus. The damage to guava fruits is limited to the skin of the fruit and there is no soft rot. The fruit spots are slightly sunken and cracked with a dark, somewhat corky aspect. When severe, the fruits appear scabbed.

A recent survey of Cephaleuros spp. host range in American Samoa found it to be extremely broad, affecting plants in dozens of families.

Fig. Typical raised, Velvety, Coppery Brown Algal Spots on Leaves of Avocado (Persea americana) caused by Cephaleuros Virescens. The Spots consist of the brightly Pigmented Thallus of the Plant Parasitic Algae.

Cephaleuros Leaf Spot of Avocado

Algal leaf spots caused by C. virescens are commonly seen on avocado. The symptoms are responsible for many of the questions raised by backyard growers. The disease usually attacks leaves on low-hanging branches.

Often, there will be no spots at all on the upper leaves. Damage to avocado plant vigour and yield is generally insignificant.

CEPHALEUROS LEAF AND FRUIT SPOTS OF GUAVA

Both C. virescens and C. parasiticus can cause substantial damage to some varieties of guava in Hawai'i. Leaf spots can be severe, leading to reduced plant vigour and defoliation. Fruit spots can affect marketability of fresh fruits and cause fruits to be rejected by harvesters or at juicing facilities.

The disease occurs commonly wherever Psidium species are growing in relatively wet conditions, such as the eastern half of the island of Hawai'i or in many of Hawai'i's coastal forests. There are two main types of symptoms on guava leaves: velvety spots and non-velvety spots. C. virescens leaf spots on guava consist of velvety patches of algal bloom on the upper leaf surface. Fruits can be severely spotted. C. parasiticus causes a nonvelvety spot type that mimics a rust disease.

Left: Cephaleuros parasiticus spots on guava leaves appear initially as tiny, dark brown specks that enlarge into roughly circular lesions with ash-coloured centers and dark brown to blackened margin. Right: On the underside of the same leaf, spots appear as dark gray, circular and watersoaked. The brightly pigmented algal thallus and sporagiophores emerge in yellowish-orange clusters within rings in the dark, watersoaked lesions. Lesions tend to form along the leaf midrib and as damage progresses the vein can collapse and die.

Fig. Cephaleuros Parasiticus Leaf Spots on Some Guava Cultivars are Surrounded by Yellow Halos, and the Associated Fruit Damage can be Significant.

INTEGRATED MANAGEMENT OF CEPHALEUROS DISEASES

Accurate diagnosis of the disease is by microscopic examination of the alga.Diseases caused by Cephaleuros spp. in Hawai'i rarely warrant management, except in the case of guava in wet environments. However, the disease may be a nuisance in landscapes.

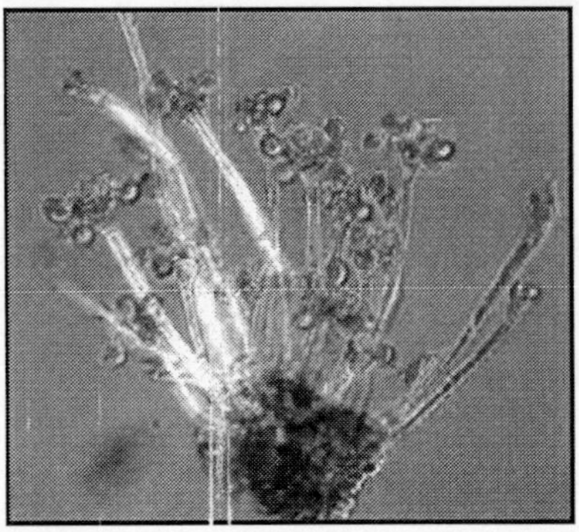

Fig. Cephaleuros parasiticus thallus and sporangiophores. Upon making a slide mount in a drop of water, zoospores may be relseased immediately

To control these parasites, integrate a combination of the following practices:

- Sanitation: Remove spotted leaves by hand from lightly diseased plants; prune low hanging branches that are affected by the disease; collect and discard all fallen leaves.
- Pruning: Prune overhanging trees around diseased plants (this will reduce relative humidity and speed up leaf drying after rains).
- Reduce plant stress: keep plants well fertilized; improve soil drainage.
- Weed control: Keep weeds around affected plants under control (this will reduce relative humidity in the plant canopy and reduce plant stress); prevent climbing and vine-like weeds from establishing on the host plant.
- Fungicides: Some fungicides will provide control; use chemicals only according to directions on the label; the Bordeaux mixture is commonly recommended.
- Intercropping: Interplant very susceptible hosts with less susceptible hosts.
- Variety selection: Try to select a more tolerant variety (for example, of guava) where the crop is grown in a wet environment.

Fig. Guava cultivars may express symptoms differently.

- Choice of planting location: For highly susceptible crops, select a planting location with moderate rainfall, well drained soil and good air circulation.
- Manage: Plant spacing and thinning to improve aeration and light exposure.

Here, leaves of three Psidium cultivars growing in a UH-CTAHR collection in the Hilo area display their unique reactions to association with Cephaleuros parasiticus.

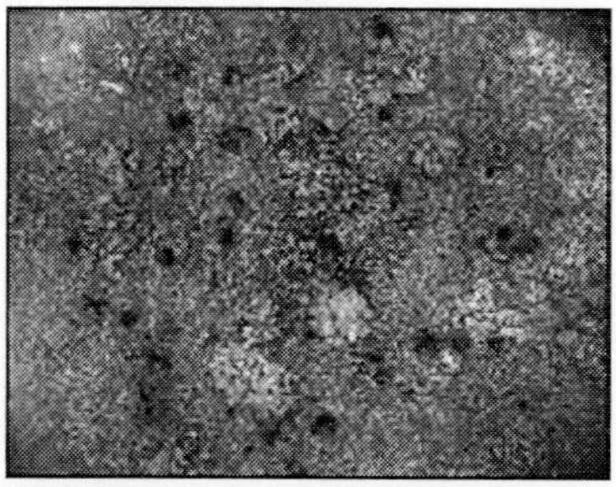

Fig. Cephaleuros parasiticus algal thallus emerging through the epidermis of the underside of a guava leaf (Psidium guajava, highly magnified).

This brightly pigmented alga is often mistaken for a rust fungus. The hematochrome pigments, astaxanthin and B-carotene, lend a yellowgreen to dark reddish-orange colour to the thallus of Cephaleuros.

ALGAL LIFE CYCLES

"Algae" is an older term used for simple, aquatic plant-like organisms that lack any sort of vascular system and are generally characterized by unicellular sexual structures. Of the various types of algae, the Chlorophyta (green algae) represent the group that served as the source for the evolution of higher land plants. This evolutionary linkage is demonstrated by two fundamental similarities:

- The fact that all the details in the complex process of photosynthesis are identical with respect to the green algae and the higher land plants. Other algal groups differ respect to the detailed types of chlorophyll and accessory pigments, the details of carbon fixation and biosynthesis, and the end-products of the process.
- The strong parallels between algal life cycles and the complex, derived life cycles of higher plants.

Some green algae are unicellular and demonstrate the simplest possible life cycles. A good example of this is provided by the life cycle of the unicellular fresh-water alga, Chlamydomonas:

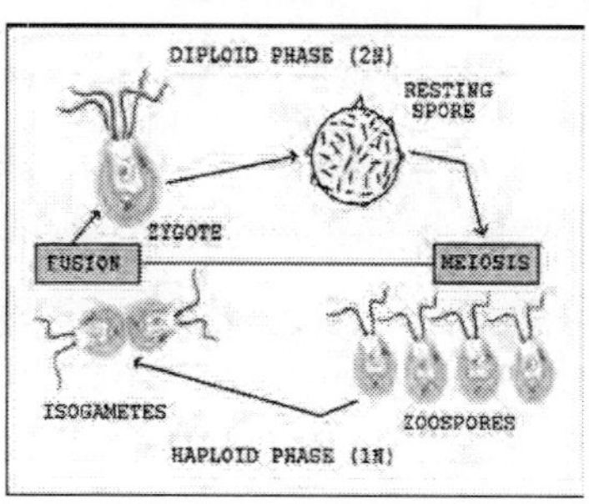

Chlamydomonas is a single-celled green alga found in freshwater habitats such as ponds and lakes. The cells swim using a pair of whip-like flagella. As long as growth conditions are good, these organisms divide asexually (binary fission) with the best-adapted genotypes making up most of the population. When conditions begin to deteriorate (the onset of winter, small ponds drying up, etc.) the individual cells each begin functioning as isogamous gametes, fusing to form motile, diploid zygotes.

The zygote almost immediately loses its four flagella and develops a very thick cell wall, forming a resting spore. This spore persists through the unfavourable period. When conditions begin to improve, each diploid resting spores divides by meiosis, producing four motile cells known as zoospores. These cells proliferate via asexual reproduction during the growing season.

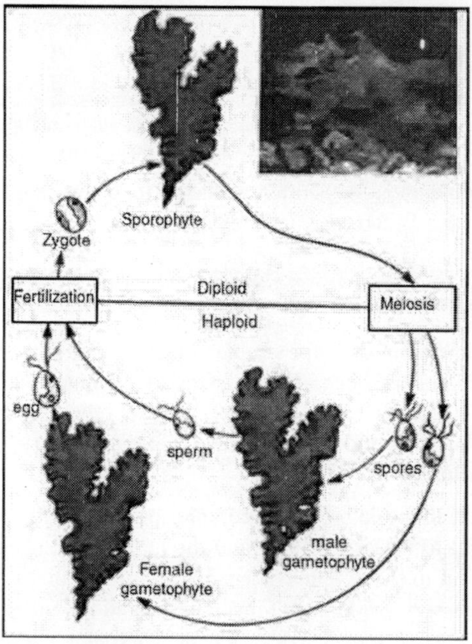

Such as a basic life cycles provides a way to preserve genetic diversity during unfavourable growth periods and then flood the environment with genetically diverse zoospores at the beginning of each growing season. In the case of more complex freshwater algae, there is a multicellular haploid stage, typically consisting of long filaments of cells which are attached at one end to the bottom or other substrate (benthic). This stage, which dominates the life cycle, is developed from the haploid zoospores produced by meiosis of the diploid resting spore. Individual cells of the filament will produce unicellular, chlamydomonas-like gametes when environmental conditions deteriorate.

- Complex plants always have a multicellular haploid (gametophyte) stage in their life cycle!

In essence, the complex diploid (sporophyte) stage of the Ulva life cycle is simple a diploid version of the haploid growth form. Individual cells of the diploid plant undergo meiosis to produce the zoospores that develop into the haploid stage of the life cycle.

❐

3.

Infection of Seed and Transmission of Seed Borne Pathogens

As an introduction we will remind some aspects of the morphology and anatomy of the seed, in order to understand in a better way the effect pathogens have in the seed and later, the plant originated from it.

THE SEED

The seed consists of three basic parts:

- Embryo,
- Storage tissues
- Seed coat.

The embryo has a reproductive function, being capable of initiating cellular divisions and growth. It is the most important part of the seed. It is an axis that originates growth in two directions, with the objective of originating a root and a shoot. Usually, the embryo is very small, compared to other parts of the seed. It is like a miniature plant. It is consists of a radicle, plumule, one or two cotyledons, and hypocotyl or epicotyl, depending on the type of plant.

Energy storage is localized in the cotyledons, in the endosperm or the perisperm. Cotyledons originate from the zygote and are part of the embryo. In many species, storage substances are localized in the cotyledons, and the embryo develops absorbing all the endosperm.

Episperm is the seed cover, consists of two layers, the testa and the endopleura. The outer layer is the testa; it can be stony, leathery, membranous or fleshy. Over the testa we can recognize: the hilum, scar or point of attachment of the seed to the funiculus, water penetrates easily through it; Micropyle, the point upon the seed at which was the orifice of the ovule through which the pollen tube enters; raphe, suture originated by the part of the funiculus that is fused along the side of the ovule. The endopleura is the inner layer; it is thin and generally whitish. Teguments, testa or protective covers delimit the seed. They are formed by one or more layers of cells originated from the ovule integuments and sometimes from the pericarp made from the walls of the ovary.

Regarding the healthiness or pathology of the seed we will consider three aspects that allow us to understand processes associated to seed-borne diseases: infection of the seed, infection of the plant and steps that can be taken to reduce the damage caused by this relationship.

Seed Infection Mechanisms

The area of science that studies the relationship between pathogens and seeds is Seed Pathology. It does not only identify the pathogens, it also includes the role of the seed as source of inoculum, the survival of the pathogen and the actions taken to control the pathogens associated to it. It uses the knowledge of General Pathology, Microbiology and Seed Analysis. To obtain a perspective of seed borne diseases, seed borne microorganisms can be considered fewer than four classes.

Fig. Infection of the Seed

The first consists of pathogens for which the seed is the main source of inoculum; when seed infection is controlled, the disease is controlled. An example would be lettuce mosaic virus. For these pathogens, the importance of seed-borne inoculum has long been recognized, and control practices have been developed. The second class consists of important pathogens in which the seed borne phase of the disease is of minor significance as a source of inoculum.

Examples are those in which the crop residues in the field were the major source of inoculum. The third and largest group of seed borne microorganisms consists of those that have never been shown to cause disease as a result of their presence on seeds. The fourth class is a group of microorganisms that can infect the seed either in the field or in storage and reduce yield and seed quality. Examples of field fungi are Diplodia, Fusarium, Cladosporium, etc.

The storage fungi Aspergillus and Penicillium can invade most types of seeds under high-moisture storage conditions. The process of seed infection is influenced by the conditions under which the crop grows.

Between the facts that influence in the process of infection are: the host and its genotype, the pathogen and its genotype end environmental facts. An infected seed will not always an infected seed will be the cause of the infection of the infection in the plant which originates, so, it is to say that the infection caused from an infected seed is an exception, not something to happen generally. There are two situations, Systemic infection of the seed and Contamination or Infestation of the seed.

Systemic Infection of the Seed

The establishment of a pathogen in any part of the seed is refereed as seed infection. It can be systemic, by the vascular system or plasmodesmata or directly by natural or artificial wounds. The same pathogen can infect the seed using one or more of these mechanisms.

For example: Xanthomonas campestris pv. phaseoli can infect seed through the vascular system, by natural openings (From the pod suture goes to the funiculus, then to the raphe and tegument, or it can also happen thought the micropyle).

Systemic Infection through Flowers, Fruits or Funiculus

Most of the systemic seed-borne bacteria and fungus reach and infect the embryo through the flower or from the peduncule of the fruit, via funiculus. Viruses go to the embryo from the systemically infected mother plant and the infected or contaminated pollen. They rarely reach the embryo during the formation of the seed or the embryo itself.

Examples of some infections that occur through the vascular system are: Some forma specialis of Fusarium oxysporum in pumpkin, pea and tomato; Plasmopara halstedii in sunflower; Septoria glycines in soy; Verticilium dahliae in spinach and sugar beet; Certain pathovares of Xanthomonas campestris in bean, cabbage, rice and sweet pepper; and Pseudomonas syringae pv. Lachymans in cucumber.

Penetration through the Stigma

During the infection, some pathogens follow the same path as the pollen grains do. Spores of some fungus reach the stigma and germinate, producing an hypha that reaches the ovary through the style, where they can stay as a dormant mycelium until seed germination.

For example: Ustialago Nuda and U. triticc in baraley and wheat, and Alternaria alternata in sweet pepper. Viruses can infect through infected pollen, the male gamete carries the virus, when joining the ovule it generates an infected embryo. If both, the male and female gamete are infected they can even produce an infected endosperm.

Penetration through the Wall of the Ovary

Some fungi, like Ustilago nuda and U. triticci penetrate through the wall of the ovary as a result of the germination of the Teliospores on the stigma or the wall of the ovary. The pro-mycelium goes through the wall and other tissues until it reaches the embryo. In some other cases, penetration occurs through breakages on the testa, establishing itself in the endopleura or the endosperm. In fleshy fruits, like cucumber, melon, eggplant, tomato, sweet pepper and others, contamination can occur directly through the funiculus or in the tegument, during the process of seed formation.

Examples of this are Colletotrichum lagenarium in watermelon and Rhizoctonia solani, when it invades fleshy fruits, it is capable of infecting from the placenta and penetrate to the developing ovule or seeds that are still in its formation process and have not lignified its cover.

Penetration through Wounds and Natural Openings

Natural openings like the hilum and the micropyle or wounds generated during the thresh are spots where pathogens like Xanthomonas campestris pv. phaseolicola in bean and Pseudomonas syringae pv. lachrymans in cucumber.

SEED CONTAMINATION OR INFESTATION

Contamination or infestation refers to the passive relationship of a pathogens and seeds. The pathogen itself or parts of it can stick or can get mixed with the seeds during any of the processes during seed recollection: harvesting, extraction, thresh, selection and packing.

Pathogens that Stick to the Surface of the Seed

Pathogens that stick to seeds during harvest or postharvest do so by their spores (Clamidospores, Oospores, Teliospores, Uredospores), bacterial cells and in some cases, virions. Spores of the following fungi can be carried on seed coat surfaces: Alternaria brassicae and A. brassicicola in crucifers; A. longipes in tobacco; A. radicina in carrot; Ascochyta pinodella in pea; Drechslera sorokiniana and D. oryzae in rice; D. avenae in oats; Fusarium oxysporum f. sp. callistephi in China aster; Sclerotia of Rhizoctonia solani in eggplant, pepper, and tomato; Tilletia caries, T. foetida, T. contraversa, and Urocystis agropyri in wheat.

A number of bacteria, such as the following, contaminate seed surfaces: Corynebacterium flaccumfacjens pv. flaccumfacjens in bean, Pseudomonas syringae pv. phaseolicola in bean, P. syringae pv. lachrymans in cucumber, P. syringae pv. tomato in tomato, C. rnichiganense pv. michiganense in pepper, Xanthomonas campestris pv. campestris in cabbage Also some viruses as Tobacco mosaic virus, Tomato mosaic virus, Pepper Mosaic virus.

ACCOMPANYING CONTAMINATION

This type of contamination refers to physical mixing of the seed with pathogen's propagation organs like the sclerotium, nematode's galls, contaminated plant parts or soil particles containing pathogens.

Structures of the Pathogens

Some fungi produce resistance structures named sclerotiums that are made of compacted mycelium. Under certain conditions of temperature and humidity it can germinate, it is common that this occurs together with the seed, producing its infection, by direct penetration or by the spores generated by the fructification organs (Apothecium or Perithecium). These spores are carried by the wind and taken to susceptible tissues of the plant, generally, some parts of the flower. Sclerotiums have many shapes and during thresh can be easily included with the rest of the seed.

Some examples of this are Sclerotinia sclerotiorum in horticultural, grain and flower crops, Sclerotinia cepivorum in garlic and Claviceps purpurea in barley.

Mix with infected Plant Parts

Infected plant's parts and residues can carry fructifications or spores of fungi and bacteria. This situation is particularly common in cereal, forage and oily crop's seeds.

Examples of this are Septoria nodorum in wheat, Puccinia malvacearum in hollyhock (malvarrosa), Colletotrichum trifolii in clover, Erwinia carotovora pv. carotovora in tobacco, Pseudomonas syringae pv. phaseolicola in bean and Sclerotinia sclerotiorum in sunflower, soy and peanut.

Soil

Seeds can be mixed with contaminated soil and, carry micro-sclerotiums of Macrophomina phaseolina in dirty seeds of kidney bean, Verticillium oxysporum f.sp. phaseoli in cotton seeds, Plasmodiophora brassicae in turnip and Fusarium oxysporum f. sp. phaseoli in bean.

SEED TRANSMITION

We saw how the seed gets infected or infested. Now we will see how these seeds can produce an ill plant. This is to say, how the inoculum goes from the seed to the plant. We call seed-borne pathogens only those that can produce an infection. Because they have to be distinguished from those that pathogens not transmitted by seed.

The virus that causes curly top in sugar beet can be present in the perisperm but does not cause an infection in the plant. In general terms, infection can be classified into systemic and non-systemic. It is systemic when the pathogen introduces itself to the plant when the seed germinates, and develops with it. Non-systemic infection occurs when there is a localized infection caused by the pathogen in the seedling at the stage of pre or post emergency, in this situation there are no systemic symptoms.

Systemic Transmition

This type of infection can be produced by pathogens that are carried with the seed in various parts, like the embryo, endosperm or episperm, or by contamination of the outer portion of this one.

Infection of the Embryo

When the seed germinates, if the embryo is infected, the pathogen initiates its growth together with the plant. Symptoms can show up during different stages of development. In the case of Ustilago triticci, the mycelium grows together with the plant and expresses symptoms only at the flowering stage, which are that all the tissues of the spike, exept the rachis, are replaced by spores.

Some pathovares of Xanthomonas campestris that infect cabbage, bean or sweet pepper move between the cells of the host until they reach the vascular system and produce symptoms in leaves or stems. Most of seed-borne viruses persist inside the embryo. Its multiplication and movement accompanies the plant during its development, and there can be symptoms at any stage, from the formation of first leaves until flowering or fructification.

Non-Embryo Infection

Infection of the episperm (testa and endopleura) occasionally conducts to a systemic infection. Some bacteria, like Corynebacterium michiganense pv. michiganense in tomato and Xanthomonas campestris pv. campestris in cabbage, penetrate through stomata of cotyledons, and from there, reach the vascular system, initiating the systemic infection.

Episperm Contamination

In some few cases this type of contamination conducts to a systemic infection. Most of these exceptions are fungi that are highly specialized in their pathogenesis and produce in cereals the so called smuts, rusts or mildews. In these cases, generally, spores are carried outside the seed, they germinate, penetrate the coleoptile, and start a systemic infection. This can occur directly o by a more complex system of haploid hypha fusion in genera like Tilletia, Ustilago, Uromyces, Sclerospora, Pernospora and Puccinia.

Non-Systemic Transmition

Non-systemic infection is very common, and in the same way as systemic infection, it can come from an infection, outside contamination or by pathogens mixed with the seeds.

Infection of the Embryo

This case is restricted to some pathogenic fungi that maintain itself in the embryo or the episperm as hypha inside the seed. Primary infection starts as injuries in the cotyledons or primary leaves, stems or petiole. Fungi fructifications (Pycnidium, Acervulus) can develop on these organs, under certain favourable conditions (temperature and humidity).

These fructifications produce spores that, with the action of water and wind, disperse the disease to other parts of the plant and other plants. This occurs in Ascochyta pisi in pea, Colletotrichum lindemuthianum in bean and C. truncatum in soy.

Infection of the Episperm

Generally, seeds in which the episperm is infected, do not geminate, or germinate and contaminate the soil. Rarely produce a systemic infection, but can infect the seedling from the outside. Over the injuries, new inoculum is produced; this one infects other parts of the plant or other plants. Septoria nodorum in wheat under high humidity conditions forms Picnidium in the coleoptile, whose spores disseminate the disease to other plants.

Contamination of the Episperm

Contamination outside the seed's testa can produce healthy seedlings, but the inoculum infects the soil and from there it can cause infections at more developed stages of the plant. Wheat and rice grains that are contaminated with Teliospores of Neo-vossia indica in wheat and N. torrida in rice, produce Esporidisporas in the soil, that can be carried by the wind and infect flowers, originating the smuts.

Accompanying Contamination

The mixture of seeds with sclerotiums of the fungi or contaminated soil particles produces non-systemic infections, at any stage of the plant growth and development. It is common for Sclerotinia sclerotiorum to produce mycelium from the sclerotiums. These mycelium can directly infect the plant or produce fructifications that produce spores that can be carried by the wind and infect flowers of different species, like sunflower, soy, peanut, etc.

Prevention of Seedborne Diseases

Prevention of seed-borne diseases is based in two main facts. On one hand there is the usage of material free from pathogens and contaminants that could originate a disease and on the other there are the treatments to eliminate the inoculum in the seed or propagule. When using material free from pathogens it is important to have adequate pathogen detection methods, this is to say, sensible and specific. No production of health quality seed can be done without valid methods to detect the possibility of contamination. Some treatments can be necessary to eliminate the pathogens from the seeds, or at leafs, maintain an economically viable balance.

Health Testing of the Seed

One of the most relevant aspects about controlled health quality seed is refereed to the detection methods used on every case. ISTA (International Seed Testing Association) appeared in 1924, by was only interested in seed's gene purity and germination aspects. But detection methods were applied.

According to the criterion of each laboratory, so their results were not comparative. So, in 1957, PDC established a comparative health testing programme and standardize applied methods. The complexity for comparison of these methods made a considerable step back on the publication of working sheets (protocols). ISTA approved methods became Rules. By 1999 only 64 protocols were being compared, and only 14 of them were accepted as Rules.

As the solutions proposed by ISTA did not satisfy the needs of international seed industry, this industry, in 1994, organized The International Seed Health Initiative for Vegetables (ISHI-Veg) chartered by the vegetable seed industries in The Netherlands and France. The Initiative was soon joined by the seed companies in the United States, Israel and Japan. This group represents the production of over 75 per cent of the worlds vegetable seed supply. ISHI published many protocols, organized by crop. For horticultural crops, the ISHI-VEG, the Seed Health Testing Methods

ISHI-Veg Status

- An ISHI Validated Reference Method is a method that has been through the ISTA/ISHI comparative testing process and is being published as ISTA wn ISHI Reference Method is a method that has been through ISHI comparative testing and is under review by the ISTA/ISHI reviewers for publication as ISTA Working Sheet.
- An ISHI Accepted Method is a method that is commonly used by the seed industry for determining seed health. The method has been published in a journal, as a working sheet and/or is publicly available for comparative testing and commonly in use by the seed industry. The method is in comparative testing or aspects of the test are being tested for inclusion or deletion from the method.
- An ISHI Reviewed Method is a method that is publicly available or published, documented and prioritized by the ITG, but not subjected to a comparative test at this time. The method usually is accepted by other agencies or groups.

In 1995, during the second ISTA-PDC symposium it was decided that a Joint ISTA/ ISHI Guidelines for Comparative Testing of Methods for Detection of Seed Borne Pathogens would be edited.

Through this combined effort many laboratories, organized under this alignments, the ISTA developed the Manual of Seed Health Testing Methods that has two sections:

- Validated Test Methods
- Peer Reviewed Methods.

The transition process from working protocol to ruled method is long and laborious, so working protocols can be considered as adequate methods for seed analysis. Nowadays, it is common that technological advances may become obsolete even before they are published.

The protocol to analyse Xanthomonas campesrtis pv.campestris, which causes black rot in crucifers, was published in 1982, there, it was recommended one minute to extract the bacteria from the seeds to starch culture medium. 2 years later, this time was raised to 2 hours, and other specific culture mediums. In 1987 the protocol was replaced, again in 1996 and at present time again it is under revision.

Minimum requirements to Produce Certified Seed

The Organization for Economic Cooperation and Development (OECD) establishes steps to be taken in order to produce basic and certified seed. Between these steps it is emphasized that generations from the mother material to the basic material are strictly limited in number. The number of certified seed generations, after the basic seed, varies from crop to crop and technical conditions.

In both, sexually and vegetative reproduced species, the schemes include similar steps. It generally starts with selection of a plant with genetic and health advantages. If health condition was not satisfying, certain actions should be taken in order to release it from considered pathogens. This Mother Plant is maintained under strict special security conditions in order to maintain healthiness (isolated glasshouses or chambers).

In the Certification Schemes two particularities abut seed borne pathogens should be considered, those transmitted exclusively with the seed and those that can also be transmitted with contaminated soil and plant rests. Diseases transmitted exclusively with the seed can be controlled by using clean healthy seed.

This seed can be obtained by Seed Certification Programmes or by treatments that are really effective, like in vitro culture, thermotherapy, chemotherapy or a combined action of these, for systemic pathogens.

Tolerance levels in certified seed depend on the multiplication speed and dispersion of the inoculum of the pathogen of the considered disease. If a small amount of inoculum may be epidemic, like Anthracnose and the bean bacterial blight, if levels above zero chemical control may not be enough to stop the disease.

But pathogens whose inoculum multiplies slowly generation after generation, the disease may be evident only several generations later, so, if contamination levels are above zero may be accepted in adequate control methods are applied to the seed. This is what happens in wheat, with the Loose Smut, continuous treatments diminish inoculum levels, and make it perfectly tolerable.

New Techniques for Seed Health Analysis

Pathogen detection methods have suffered great changes since 1957. Back then it was mainly targeted to fungi, and based exclusively on incubation and identification, or the cultivation of these seeds and counting the percentage of unhealthy seedlings. These analyses were very laborious and required a high infection level in order to be detected.

Nowadays, technological advances have allowed detecting even low infection levels, evaluating 1,000 to 10,000 seeds at once. Specificity of culture mediums, usage of antibiotics and other products has allowed isolating the pathogen from the seed.

Together with this, the use of reactives as mono or polyclonal antiserums and molecular markers is also available. The Lettuce Mosaic Virus (LMV), back in 1950 the symptoms were observed over many thousands of seedlings from the seed that were going to be analyzed. Later, technical advances determined that 30,000 seedlings were needed, and even later, in 1983, an alternative to the Chenopodium quinoa test was discovered, the ELISA test.

Watermelon Fruit Blotch caused by bacteria is a severe problem in glasshouse winter production. It was determined that a single contaminated seed in 9,000 of them was enough too widespread the bacteria to the rest of the plants. That is the reason why screening was done over 10,000 seedlings. Nowadays PCR technique is used.

❐

4.

Phytoplasmas and Spiroplasmas

PHYTOPLASMAS

Phytoplasmas are prokaryotes lacking cell walls that are currently classified in the class Mollicutes. Phytoplasmas are associated with plant diseases, and are known to cause more than 600 diseases in several hundred plant species.

The symptoms shown by infected plants include: yellowing or reddening of the leaves, shortening of the internodes with stunted growth, smaller leaves, excessive proliferation of shoots resulting in a witches' broom, phyllody, virescence, sterile flowers, necrosis of the phloem tissues, dieback of the branches of woody plants, and the general decline and death of the plant.

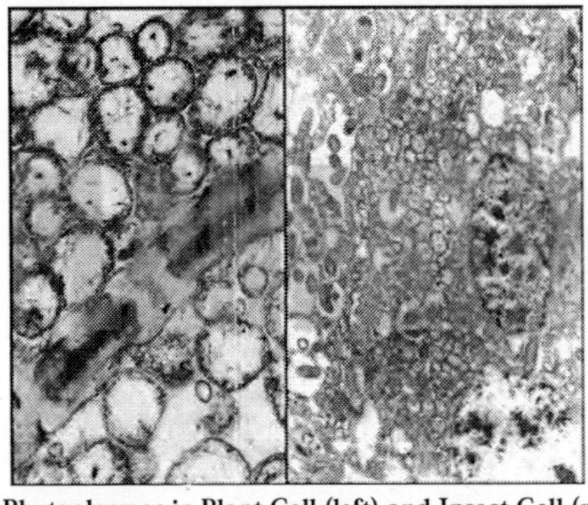

Fig. Phytoplasmas in Plant Cell (left) and Insect Cell (right)

Phytoplasmas are transmitted from plant to plant by insect vectors, mainly leafhoppers and psyllids (Ploaie, 1981). They traverse the wall of the intestinal tract, multiply in the hemolymph, and pass through the salivary glands, in which they multiply further. Then, the insect vectors introduce phytoplasmas along with salivary fluids into the phloem of a new host plant. Usually these insect vectors do not transmit phytoplasmas transovarially, although two exceptions have been reported: aster yellows and mulberry dwarf phytoplasmas.

Morphology

Being mollicutes, phytoplasmas lack cell walls and instead are bound by a triple layered membrane.The cell membranes of all phytoplasmas studied so far usually contain a single immunodominant protein (of unknown function) that makes up the majority of the protein content of the cell membrane. The typical phytoplasma exhibits a pleiomorphic or filamentous shape and is less than 1 micrometer in diameter. Like other prokaryotes, DNA is free in the cytoplasm. They are believed to reproduce through binary fission.

Life Cycle

Because of the difficulty culturing phytoplasmas in vitro, the organisms remain one of the most poorly characterized groups of plant pathogens. In the last decade, however, molecular biological methods have made it possible to determine the phylogenetic and taxonomic relationships between phytoplasma strains, and those between phytoplasmas and other prokaryotes. Currently, the classification of phytoplasmas is mainly based on the nucleotide sequence of the 16S rRNA gene.

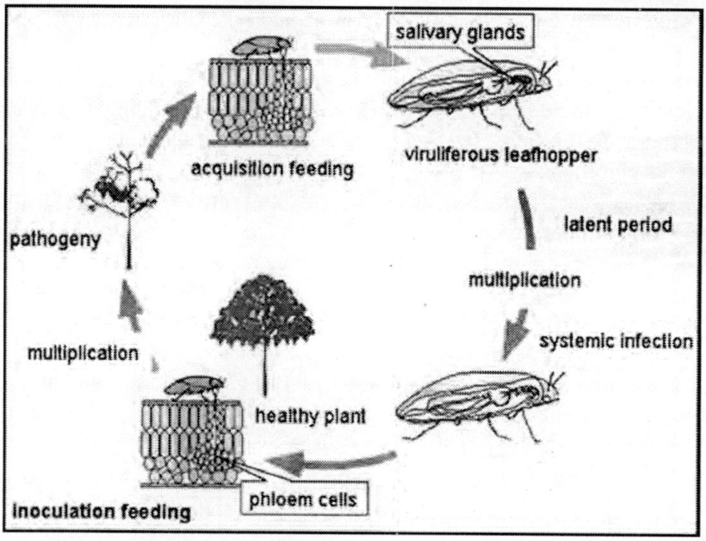

Diseases Caused by Phytoplasmas

Until recently, plant diseases known as the "yellows diseases" were thought to be caused by viruses. This was because the yellows disease agents had certain characteristics in common with virus agents. They couldn't be grown on culture media like most fungal or bacterial pathogens could. In addition, while a yellows disease had characteristic symptoms, there were never signs – obvious structures of a disease organism – on infected plants. Then, in 1967, using an electron microscope, researchers in Japan found wall- less microorganisms in yellows-diseased plants.

This new class of plant disease agent was termed a "mycoplasma-like organism," because of its resemblance to mycoplasmas that were saprophytic or that caused disease in humans and animals. In 1994, characterization of the organisms associated with yellows diseases had progressed to a point that it becam clear they were unique and should be given their own name.

The name chosen for the group was phytoplasma. An ancient group of organisms that had caused plant disease problems throughout history had a new name.

Hosts and Distribution: Diseases caused by phytoplasmas occur worldwide in many crops, although individual phytoplasmas may be limited in their host range or distribution. There are more than 300 distinct plant diseases attributed to phytoplasmas, affecting hundreds of plant genera.

Many of the economically important diseases are those of tree crops, including coconut lethal yellowing, peach X-disease, and apple proliferation.

There are also major diseases of vegetables, such as tomato big bud, and aster yellows in carrot and onion. Hosts for aster yellows phytoplasma include aster, chrysanthemum, gloxinia, monarda, and many other perennials and flowering plants.

Some phytoplasma diseases, such as elm yellows and ash yellows, are responsible for decline of forest areas, woodlots, or urban landscapes.

Disease Agent and Detection Methods: Phytoplasmas are single-celled organisms that are similar to bacteria but lack a rigid cell wall. Phytoplasmas are obligate parasites – they cannot survive apart from a host. They grow and reproduce in the cytoplasm of host cells, both in insect vectors and in plants.

Phytoplasmas are very small – submicroscopic – requiring the resolution of an electron microscope to be seen. They reproduce asexually, by budding. Phytoplasmas look like amorphous sacks or blobs, ranging from 70-1000 nm in diameter, or roughly the size of a plant cell's chloroplasts.

There are at least 75 distinct phytoplasmas, currently clustered in 14 groups and 38 subgroups, that affect several hundred plant genera. Naming of diseases and of their phytoplasma agents is currently disorganized and inconsistent.

Because the diseases were described long before the pathogens causing them could be seen or characterized, disease names are more familiar to us than pathogen names. However, as it became possible to characterize the disease agents at a molecular level, it became clear that in some cases there were several different phytoplasmas associated with a single disease syndrome. In other cases, there were several different diseases named that were all caused by one phytoplasma.

Because microorganisms require culturing to be classified properly, standard taxonomic schemes are not possible with phytoplasmas at this time. Nomenclature is still in transition; it will take more work and more time before the picture becomes clear enough to assign proper names to diseases and pathogens. This work is extremely important if we are to know how to best manage phytoplasma diseases.

Although phytoplasmas were first detected with electron microscopy, staining, serological, and DNA-based techniques have been developed that make detection more rapid and practical for diagnostic labs. Serological and DNA-based techniques provide the ability to distinguish among different phytoplasmas and permit specific identification and classification.

Transmission: Phytoplasmas reside in the phloem tissue of the plants they infect, invading primarily phloem sieve tube elements. Phytoplasmas are transmitted by phloem-feeding insect vectors, most commonly leafhoppers but also some psyllids and planthoppers. Not all vectoring insects can transmit all phytoplasmas – there are specific interactions of a particular phytoplasma with its insect vector. Some phytoplasmas, such as peach X-disease phytoplasma, may be transmitted by several species of leafhoppers; others, such as elm yellows phytoplasma, appear to be transmitted by one or only a few species. An insect vector acquires the phytoplasma after feeding on an infected plant for several hours or days. For 10-45 days, the phytoplasma moves through the insect and multiplies within specific organs.

After that incubation period, the insect is able to transmit the phytoplasma to uninfected plants when it feeds. The incubation time within the insect depends upon several factors, including temperature. An infected insect will be able to spread disease for the rest of its life. It was generally believed that phytoplasmas could not be passed from adults into eggs, but that each insect generation must re-acquire the phytoplasma from infected plants.

A recent report brings this belief into question again, and more study is needed to make certain that the lack of transmission from adult to egg is always true.

Plant host range for a phytoplasma is dependent upon vector specificity and feeding habits. Some leafhoppers are mono - or oligophagous, which means they only feed on one or a few host species. If a phytoplasma is dependent on natural spread by this vector, then it will only be spread to those species of plants fed upon by the leafhopper. American elm yellows disease, for example, has only been found in nature infecting American elm species.

It is possible to transmit the phytoplasma to European or Asian elm species by grafting infected material to those species, and some of those species develop disease symptoms.

But since the leafhopper vector for American elm yellows does not feed on European or Asian elm, it has not spread in those elms, even in areas where American elms are dying from phytoplasma disease.

In other cases, it is the phytoplasma-plant interaction that limits the host range of disease: the leafhopper may feed on an uninfected plant, but it is not a plant species that is sensitive to the phytoplasma, and the phytoplasma will not multiply or produce symptoms in the plant. Phytoplasmas cannot be transmitted mechanically, by rubbing sap from an infected plant onto a healthy plant or by using contaminated tools.

But they can infect healthy plants if infected stock is budded or grafted directly onto it. Plants growing close together can be infected through naturally-formed root grafts. Phytoplasmas are not known to be transmitted through seed or pollen.

Symptoms: Symptoms of the group include a general yellowing or reddening of leaves, reduced leaf size, shortening of internodes, and loss of apical dominance. These effects lead to stunting of the plant, proliferation of shoots or roots, witches-brooming, reduced yields, general decline, and sometimes death of the plant. Several symptoms fairly unique to phytoplasma diseases involve flowers: virescence (greening of the flowers), phyllody (conversion of petals and sepals to more leaf-like structures) and sterility of flowers. Obviously, these types of symptoms are disastrous for seed or ornamental crops.

Fig. On the Left, a Normal Onion Flower; on the Right, the Flower of an Onion Infected with Aster Yellows Phytoplasma.

Fig. Macrosteles Fascifrons, Leafhopper Vector of Aster Yellows Phytoplasma. Immature (left) and Mature Adult (right). Adults Average 3.8 mm Long.

Management and Control: Although phytoplasma disease symptoms may be delayed or alleviated by treatment with certain classes of antibiotics, this approach is not usually practical. Plants are not really cured, and symptoms reappear when treatments cease. Foliar applications or soil drenches of antibiotics are not effective. For trees, injections into the trunk are most effective. Infected plants or dormant propagative tissue can be freed of phytoplasma by heat treatment.

This technique is used at quarantine facilities when phytoplasma infection is suspected in fruit tree nursery stock being imported into a country. The most promising strategy for avoiding phytoplasma disease is the identification or development of resistant plant varieties. Management and control centre around clean stock programmes, eliminating sources of the phytoplasma, and controlling leafhopper vectors:

- Propagate from seed, or from phytoplasma-free plants.
- Eliminate perennial and biennial weed hosts.
- Rogue out and destroy symptomatic plants.
- Avoid planting susceptible crops next to crops harbouring the phytoplasma.
- Control the leafhopper vector in the crop and nearby weeds early in the season.
- Plant varieties that are more resistant to the disease, if available.

In a landscape situation, there is no need to remove diseased plants unless they are declining. Remove specimens showing progressive decline.

Phytoplasma ecology is complex, is affected by the host range and geographic distribution of both the phytoplasma and the insects that transmit them, and is strongly affected by weather conditions. As more is learned about relationships among disease organisms, vectors, and hosts, surprising ecological niches have been uncovered, and theories of how disease may have evolved have been developed. These suggestions may be important in choosing management strategies for disease, and in forecasting where new disease outbreaks may occur.

SPIROPLASMAS

Spiroplasmas, are present in the hemolymph of almost all insects; there probably are several million strains. They can also cause diseases in plants but are usually associated with a vector. For example, a leaf hopper carries a spiroplasma that infects orange trees.

Spiroplasmas are similar to mycoplasmas. They do not have a cell wall (cell wall deficient) and have among the smallest genomes of any living organism. Mycoplasma, are the smallest and perhaps the oldest life forms. These bacteria, one cause of "walking pneumonia", are thought by many to be rather fragile, but nothing could be further from the truth. They tolerate extreme fluctuations in temperature, lay dormant in the soil for generations and survive the harshest elements; only drano-like chemicals kill them effectively outside the body. Under normal circumstances our immune system efficiently deals with this complicated, membrane enclosed piece of DNA .

Spiroplasmas as the cause of CJD

A common phenomenon among the mycoplasmas is that the organisms bind host proteins that often are of identical molecular weight to their surface proteins and, therefore, are looked at by the immune system as being the same as the host.

The spiralin protein on the surface of spiroplasmas shows a migration pattern on gel electrophoresis with a molecular weight of 27,000 Da to 30,000

Da, similar to that of the prion protein. This biochemical similarity is compatible with spiroplasma etiology.

Bastian was able to show that spiroplasmas were neurotropic. When inoculated peripherally into suckling rats, they will eventually localize to the brain tissues. The organisms will produce a persistent infection and produce a spongy change in the brain tissue of these animals. The neuropathologic changes are similar to those seen in CJD.

Spiroplasmas are also within the size range of the agent that transmits CJD and other transmissible Spongiform encephalopathies. Spiroplasmas will pass through a 50 nm-pore filter. The transmissible agent's size has been determined to be 42 nm.

The obvious way to look for an agent directly is by electron microscopy, but this method may not be appropriate for spiroplasmas. Spiroplasmas are similar to mycoplasmas, and it is a well-known phenomenon that mycoplasmas are able to blend with cell membranes. What happens, possibly, is that spiroplasmas essentially fuse with host-cell organelle membranes, thereby blending with the background, so you would not see it unless you had a marker to label it. Developing such as marker has been difficult because spiroplasmas are very fastidious (difficult to cultivate)organisms.

Bastian also inoculated suckling rats with spiroplasmas, and examined their brain tissues by electron microscopy early in the infection process; he documented the organisms in the tissues. They appeared as membrane-bound forms, except for the one instance where he observed the spiral form. Later in infection, when he knew that the tissues were infectious by broth culture, he couldn't find any evidence of spiroplasmas by looking at the tissues extensively with electron microscopy.

Bastian insists that the infection-related protein that most researchers refer to as a "prion" is produced by the host in response to the infection and is not the causative agent. Prions are thought to be self-replicating proteins. Some researchers believe prions are the cause of CJD and related illnesses because they have found prions in brain tissue from people with CJD and sheep with scrapie but not in normal brain tissue. Bastian states that a shortcoming in the prion theory is that CJD and scrapie can be transmitted without prions.

Brain material from which the prion has been removed with antibodies can still infect animals. Moreover, the prion has been found in unrelated disease processes, such as Kawsaski syndrome and inclusion body myositis. Prion researchers have jumped to conclusions and have not considered any other possibility. It is quite possible that spiroplasmas may be inducing the formation of the prion protein to protect itself from the immune system.

The immune system is very important in the pathogenesis of CJD. The agent replicates in the spleen and lymph nodes and occasionally causes an immunologic reaction. Auto-antibodies are characteristically seen in the late stages of experimental and naturally occurring disease.

The gene for the host protein is located on the chromosome in the region of the major histocompatibility complex (MHC) in the mouse. "Occasionally, you see elevation of immuno globulins; there are morphological alterations of the leukocytes; there is leukopenia," Bastian explains, "and auto antibodies are characteristically seen in the late stages of both experimental and naturally occurring infection. There is partial MHC restriction in both human and animal disease." "The immune reaction seen in these Spongiform diseases can be explained by super antigen activity, Bastian says.

He notes that, normally, an antigen is presented to the cell surface in the MHC and interacts with the T-cell receptor—the antigen lying in a groove in the T-cell-MHC sets in motion the standard reaction. A super antigen, on the other hand, binds outside the groove of the T-cell and interacts with the MHC. This results in some immunoglobulin production, but only transiently. The major effect is clonal deletion of T cells, resulting in a state of immune tolerance. Autoantibodies can also form.

In Spongiform diseases, PrP presumably acts as a super antigen. It is noteworthy that inclusion body myositis, a condition in which prions are seen, is an established super antigen disease."

Dr. Bastian also notes that investigators have reported transmitting a TSE to mice from hay mites gathered from farms in Iceland where scrapie is endemic. He is virtually certain that these hay mites contain spiroplasma, noting that the investigators have not so far found PrP in the mites.

If hay mites can cause TSE, why couldn't the same be true for the maggots and mites on the Fore corpses? Could Gajdusek have overlooked the main factor connecting Kuru to the Fore women and children? Was the initial cause of Kuru the ingesting of large quantities of maggots and mites by protein famished women and children? We know that the maggots and mites contain spiroplasma. By eating the brains of the Papuans that died from Kuru, the disease(Spiroplasma) was retransmitted to those remaining, in a deadly cycle. Transmissible Spongiform Encephalopathies will continue to be misunderstood unless we begin to study and understand these simple connections.

❐

5.

Photoperiodic of Flowering Plants

FLOWERING PLANTS

Flowering plants are either classified as short- or long-day flowering. While a number of factors influence a plant's florescence, the length of dark photoperiods—also referred to as nights—trigger flowering. Short-day plants flower when the night is longer than a critical length. These plants cannot flower under long days or if artificial light is shone on the plant interrupting the "night." Short-day plants require a consolidated period of darkness before floral development can begin. Natural nighttime light is not bright enough nor does it endure long enough to interrupt flowering. While the length of the dark period required to induce flowering differs among species and varieties varietals, short-day plants generally flower as days grow shorter. This flowering period coincides with the time after the summer solstice in the Northern Hemisphere, which is during summer or fall.

Photoperiod affects flowering when the shoot is induced to produce floral buds instead of leaves and lateral buds. Note that some species must pass through a "juvenile" period during which they cannot be induced to flower. Common cocklebur is an example of a plant species with a remarkably short period of juvenility, and plants can be induced to flower when quite small. Long-day plants require shorter periods of darkness than short-day plants to induce flowering. These plants typically flower in the Northern Hemisphere in late spring or early summer.

Day-neutral plants, such as cucumbers, roses and tomatoes, do not initiate flowering based on photoperiodism at all; they flower regardless of the night length. They may initiate flowering after attaining a certain overall developmental stage or age or in response to alternative environmental stimuli, such as vernalisation (a period of low temperature), rather than in response to photoperiod.

Photoperiodism explains why some plant species can be grown only in a certain latitude. Experiments with the cocklebur have shown that the term "short-day" is a misnomer. In order to flower, cockleburs need a long dark photoperiod (of at least 8.5 hours). Interruption of the dark photoperiod prevents flowering unless it is followed by irradiation with far red (730 nm) light. Exposure to far-red light at the start of the night reduces the dark requirement by 2 hours. While commercial growers have had success in altering photoperiods to force plants to flower, for many plants, flowering is much more complex than this. Some require the correct photoperiod for several days before they will respond. Additionally, some not only require the correct photoperiod but also other environmental cues.

PHOTOPERIODISM

The reproductive cycles of many organisms, both plant and animal, are regulated by the length of the light and dark period, called the photoperiod. In flowering plants (angiosperms), flowers are organs for sexual reproduction, and photoperiodism refers to the process by which these plants flower in response to the relative lengths of day and night. Along the equator, the lengths of day and night remain constant because the sun rises and sets at the same time throughout the year.

The lengths of the day and night are also equal (each is six months long) at the exact North and South Poles due to the fact that the sun remains below the horizon for six months each year and above the horizon for the other six months. Any place else in the world, the days become longer in the summer and shorter in the winter.

The reproductive cycles of many organisms, both plant and animal, are regulated by the length of the light and dark period, called the photoperiod. In flowering plants (angiosperms), flowers are organs for sexual reproduction, and photoperiodism refers to the process by which these plants flower in response to the relative lengths of day and night.

The synchronization of reproduction with seasonal time is a very important aspect of plant physiology. Reproduction in many angiosperms is dependent on cross-pollination, the process of pollen being transferred from one flower to another. Hence, it is important for all of the plants of the same species in a given region to flower at the same time.

TRANSITION AND VEGETATIVE OF PLANT REPRODUCTIVE DEVELOPMENT

The rotation of our planet results in regular changes in environmental cues such as day length and temperature, and organisms have evolved a molecular oscillator that allows them to anticipate these changes and adapt their development accordingly. In many plants, the transition from vegetative to reproductive growth is controlled by photoperiod, which synchronizes flowering with favourable seasons of the year. One of the most important developmental decisions in a plant's life is when to switch from vegetative to floral (reproductive) growth. If flowering is initiated at the wrong time of the year, it will affect the number of seeds produced and significantly reduce reproductive success. In addition to its role in directly initiating flowering, light also serves as an entraining signal in resetting a plant's circadian clock. It is likely that all five phytochromes regulate red/ far-red entrainment signals. The circadian system is crucial for photoperiod perception, as alterations in clock function alter the induction of flowering time. In fact, many genes that regulate clock function were originally isolated as floral-timing mutants. From work on toc1, cca1 and lhy mutant lines, a molecular model for the central oscillator was proposed. In this model, TOC1 expression in the night activates CCA1 and LHY; the resulting CCA1/LHY expression in the morning represses TOC1 expression.

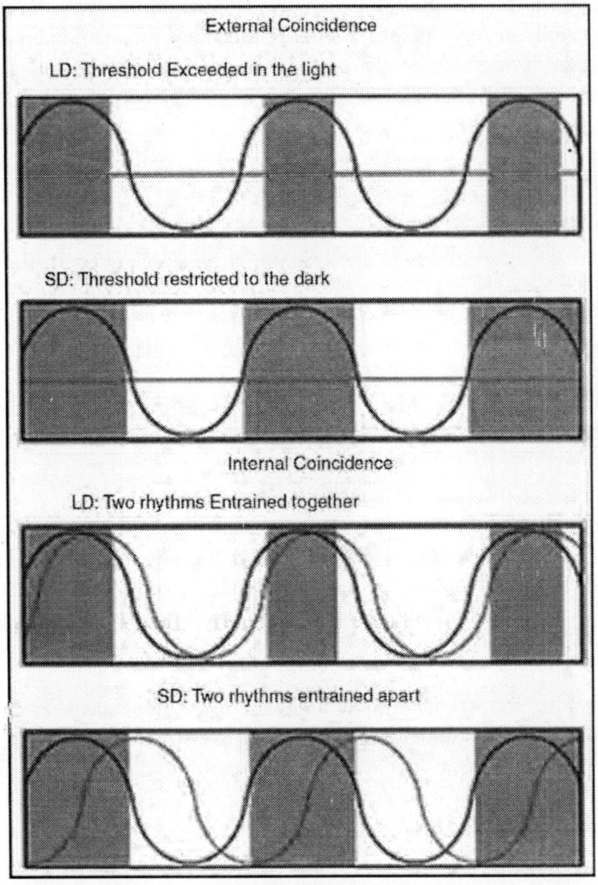

Fig. Two Models for Photoperiodic Timing

This repression is relieved by decay in CCA1/LHY expression, so that TOC1 levels increase again in the subsequent night, closing the 24 hour loop. New data argue against this model, however, as it appears that LHY and CCA1 are not absolutely required for circadian function. This has led to an alternative model, in which sequential periodic expression of TOC1 and its four homologs generates the oscillator. Regardless of the exact molecular details of the oscillator, it is clear that the circadian system integrates photoperiod Perception and the UV/blue-light-absorbing cryptochromes. Mutations in the red-light receptor phytochrome B result in early flowering, particularly in the absence of an inductive photoperiod. This establishes this receptor as a repressor of flowering time. Mutations in cryptochrome 2 cause delayed flowering time under inductive photoperiods.

Thus, cryptochrome 2 is a promotive receptor of flowering time. The competition between these two receptors could act in the determina-tion of repression or activation of reproductive development. Molecular genetic analyses have identified photoreceptors and light-signalling components, and components of the circadian system, which are essential for a plant to make a proper photoperiodic response.

Independent studies by the Carre and Kay groups have now provided a molecular foundation for understanding how light perception is integrated with the circadian system in the generation of a photoperiodic response. Their work supports the classical 'external coincidence' model which functions through the circadian expression of the CONSTANS gene and light activation of the encoded protein.

In the external-coincidence model, a physiological response, such as flowering, is triggered when light perception coincides with the time when expression of a circadian-regulated gene exceeds a required threshold. The long days (LD), light is perceived both at dawn and dusk of the expression phase, whereas under short days (SD), threshold expression is restricted to the dark. In the internal-coincidence model the effect of light is simply to entrain two distinct circadian oscillators.

In the example shown, long days cause the two rhythms to be entrained with similar phases; this could generate two regulatory molecules that require each other's activity for physiological function. Under short days, the phases of the two entrained rhythms are further apart; this could restrict the simultaneous expression of two factors, thus inhibiting their co-action.

The link in Arabidopsis between light perception, the oscillator and flowering time appears to be the transcription factor CONSTANS (CO). Loss- of-function co mutants flower late in inductive long days, whereas ectopic overexpression of CO promotes early flowering independent of daylength.

Further, CO expression level is reduced in late-flowering gi and lhy mutants, and elevated in early-flowering elf3 and elf4 mutants. Defects in CO expression can thus explain the opposite effects on flowering time of these mutations. CO expression in Arabidopsis is modulated by light perception and the circadian oscillator, and so is dynamically regulated by daylength.

In long-day photoperiods, CO abundance is high at the beginning and end of the photoperiod, but in short days, CO is restricted to the dark phase of the day. It was thus proposed that CO expression induces flowering as a function of daylength via a light-dependent post-transcriptional process that requires a threshold level of circadian transcribed CO. Further support for this view comes from the finding that the expression of the CO target gene FLOWERING LOCUS T (FT) is restricted to when the expression of CO is coincident with light. Although this supports an external-coincidence model for flowering time, only the requirement of light was directly examined in these experiments

When to flower is therefore a critical decision, and consequently multiple mechanisms have evolved to align flowering with optimal environmental conditions. But how does a plant recognize the presence of favourable conditions and integrate this information with its own endogenous developmental programme? A new clue to this problem comes from the work of He et al. 2003. To dissect the molecular processes that initiate flowering and trigger the change from vegetative to reproductive growth, biologists have carried out intensive genetic studies of flowering time in the model plant Arabidopsi. This has led to the discovery of many genes involved in the regulation of flowering time and the development of a number of genetic models.

GENETIC PATHWAYS

Despite progress in identifying the genetic pathways involved, the mechanisms by which these flowering gene products modulate the floral transition is largely unknown. He et al. in 2003 elucidated one mechanism by which plants regulate flowering time. They have identfied a protein called FLOWERING LOCUS D (FLD), which removes acetyl groups from (deacetylates) histone proteins in chromatin containing the FLOWERING LOCUS C (FLC) gene. The FLC protein encoded by this gene is a member of the MADS-domain family of transcription regulators and is a strong repressor of flowering. By deacetylating histones in FLC chromatin, FLD prevents transcription of FLC enabling the plant to flower. To ensure that flowering occurs at the correct time in Arabidopsis, floral initiation is regulated by the integration of signals from developmental pathways and environmental cues. These include pathways that monitor plant hormones, detect light quality and duration, and respond to prolonged exposure to cold (vernalization). A number of these pathways converge on the common target FLC. The autonomous promotion pathway regulates flowering independently of photoperiod (hence its name) by repression of FLC expression. FLD is one of six identified members of the autonomous pathway. Like other mutants in this pathway, fld mutants exhibit delayed flowering due to an increase in production of the flowering repressor FLC. In their new work, He and colleagues show that FLD is homologous to the human protein KIAA060, a component of the human Histone Deacetylase 1,2 (HDAC 1/2) complex. HDAC complexes remodel chromatin by removing acetyl groups from lysine residues in the tails of histones. Hyperacetylated histones are associated with transcriptionally active genes, and hypoacetylated histones with transcriptionally silent chromosomal regions. To examine whether FLD may be a component of a plant HDAC complex, He and co-workers studied the acetylation state of histone H4 at the FLC locus. They immunoprecipitated specific chromatin fractions with an antibody against acetylated H4 histone tails and inspected the different fractions using polymerase chain reaction—a technique called chromatin immunoprecipitation (ChIP). They found thatin fld mutants of Arabidopsis, histone H4 tails in a specific region of the FLC locus are hyperacetylated compared with those in wild-type plants.

So, when the deacetylating activity of FLD is disrupted, FLC remains acetylated and is actively transcribed, resulting in a delay in flowering. HDACs are often recruited to their targets through interactions with cis-regulatory elements. To define these cis-elements, He and co-workers generated a series of internal deletions of the FLC gene and introduced them into plants lacking functional FLC. Removal of a 294–base pair (bp) region within intron 1 of FLC promoted hyperacetylation of FLC chromatin, FLC expression, and late flowering. Thus, deletion of this specific region of intron 1 has effects on FLC expression similar to those observed in fld mutant plants. Taken together these results show that FLD regulates FLC by deacetylating histones in FLC chromatin, providing evidence for chromatin regulation of FLC expression. The other components of the autonomous pathway also down-regulate FLC expression.

But He et al. demonstrate that, with the exception of FVE, they do not cause similar acetylation changes at the FLC locus. The autonomous pathway, therefore, must exploit multiple mechanisms to regulate FLC. Other components of the autonomous pathway include LD, which encodes a homeodomain protein; FPA and FCA, which encode RNA-binding proteins; and FY, which encodes a polyadenylation factor.

EXACT MECHANISM

However, the exact mechanism by which these components regulate FLC is still not known. Further studies examining how FLD regulation of FLC is integrated with information from other pathways modulating FLC will provide a useful model for understanding how regulatory networks converge on a single target. Although the homology of FLD to a component of the mammalian HDAC1/2 complex would imply that FLD is part of a similar complex in plants, this still needs to be demonstrated. Four genes with homology to HDAC1/2 have been found in the Arabidopsis genome. However, the effects of mutations in these genes do not resemble those in fld mutants.

This may reflect redundancy among the HDACs such that no one mutation alters flowering time. Consistent with this, treatment with an antisense transgene that is likely to suppress several of the HDACs does result in late flowering. The discovery of multiple FLD homologs in Arabidopsis raises the possibility that different FLD-like proteins are part of HDAC complexes with different target-site specificities. The He et al. study encourages further analysis of the presumptive FLD complex, which should provide a greater understanding of the evolutionary conservation of its constituent proteins and their involvement in transcriptional repression in both plants and mammals. Such an analysis may also reveal the ways in which plant development is more plastic and adaptable than animal development. Many plant cells are totipotent, so plants need versatile mechanisms to reprogram chromatin states. Many tools are now in hand to dissect these mechanisms and to address how they differ from those controlling reprogramming of animal genomes.

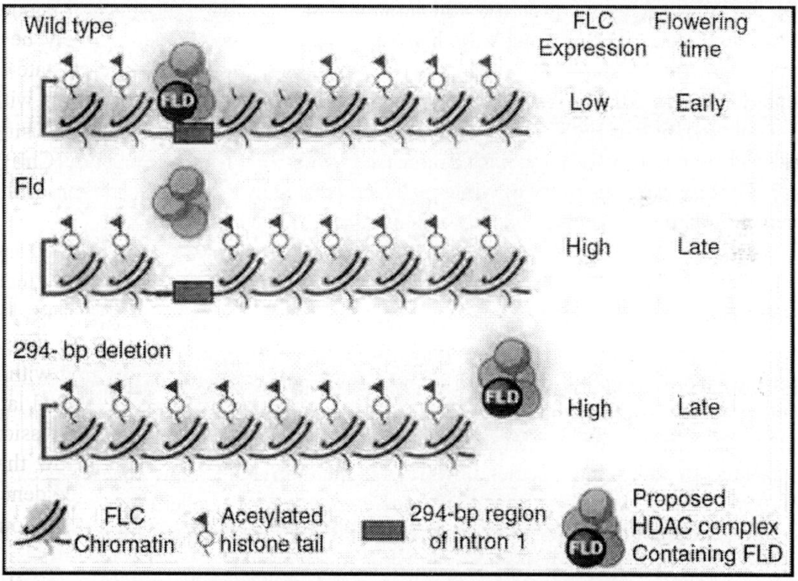

Fig. Chromatin Regulation and flowering. Regulation of theFloral Repressor FLC by FLD.

The acetylation states of FLC chromatin in wild-type plants (top), fld mutants (middle), and plants lacking a specific 294-bp region of FLC intron 1 (bottom). (Top) In wild-type plants, FLD acts as part of an HDAC complex, which interacts directly or indirectly with intron 1 of FLC to deacetylate specific regions of its chromatin. Consequently, FLC expression is reduced and the plant is able to flower. (Middle) In plants lacking functional FLD, FLD- dependent deacetylation of FLC is lost. Presumably, the HDAC complex containing mutant FLD can no longer be targeted to FLC. Therefore, the FLC locus remains acetylated, is actively transcribed, and flowering is delayed. (Bottom) Removal of a 294- bp region of FLC intron 1 also prevents FLD- dependent deacetylation of this locus possibly because the HDAC complex can no longer bind to FLC. Thus, FLC remains acetylated, and the gene is expressed, with a consequent delay in flowering.

In winter-annual types of Arabidopsis, flowering is delayed unless plants are vernalized. The delayed flowering is due to dominant alleles of FRIGIDA (FRI) and FLC. FRI elevates expression of the MADS-box transcriptional regulator FLC to levels that suppress flowering. Vernalization promotes flowering primarily by repressing FLC expression. The repressed state of FLC is maintained through mitotic cell divisions after a return to warm growing conditions. Many summer-annual accessions of Arabidopsis flower rapidly without vernalization because such accessions lack an active FRI allele or have a weak FLC allele and thus have low levels of FLC expression.

To ensure flowering in favourable conditions, many plants flower only after an extended period of cold, namely winter. In Arabidopsis, the acceleration of flowering by prolonged cold, a process called vernalization, involves downregulation of the protein FLC, which would otherwise prevent flowering. This lowered FLC expression is maintained through subsequent development by the activity of VERNALIZATION (VRN) genes. VRN1 encodes a DNA binding protein whereas VRN2 encodes a homologue of one of the Polycomb group proteins, which maintain the silencing of genes during animal development. Here we show that vernalization causes changes in histone methylation in discrete domains within the FLC locus, increasing dimethylation of lysines 9 and 27 on histone H3. Such modifications identify silenced chromatin states in Drosophila and human cells. Dimethylation of H3 K27 was lost only in vrn2 mutants, but dimethylation of H3 K9 was absent from both VRN1 and VRN2, consistent with VRN1 functioning downstream of VRN2. The epigenetic memory of winter is thus mediated by a 'histone code' that specifies a silent chromatin state conserved between animals and plants.

PHOTOPERIODIC CLASSIFICATION OF PLANTS

The photoperiodic classification of plants is usually made on the basis of flowering, but other aspects of their development may also be affected by day length. Based on their flowering response, plants are classified as short-day plants (SDPs), long-day plants (LDPs), intermediate-day plants, ambiphotoperiodic, or day-neutral plants. Short-day plants flower when the days are relatively short (generally nine hours or less), such as in the late fall or early winter. In some SDP species flowering is qualitative, meaning that short days are absolutely required, while in other SDP species flowering is quantitative, which means flowering is accelerated under short days, but short days are not an absolute requirement. Some examples of SDPs include rice, cocklebur, and soybean.

Long-day plants flower when the days are relatively long (generally fifteen hours or greater), as would occur in late spring and early summer. As with SDPs, there are qualitative and quantitative species of LDPs. Intermediate- day plants require quite narrow day lengths (between twelve and fourteen hours) in order to flower, and flowering is inhibited by either short or long days. Sugarcane is an example of an intermediate-day plant.

Ambiphotoperiodic plants are a specialized group of plants that will flower in either short days or long days, but flowering is inhibited by intermediate day lengths. In day-neutral plants, flowering is not regulated by day lengths. In other words, day-neutral plants flower regardless of the day length. There are also many interesting interactions between photoperiod and temperature.

A plant may respond to a certain day length at one temperature but exhibit a different response at another temperature. For example, both the poinsettia and morning glory are absolute SDPs at high temperature; however, they are absolute LDPs at low temperature and day-neutral at intermediate temperatures.

CHEMICAL CONTROL

Flowering is regulated by chemicals produced in the plant, and a variety of plant hormones, including auxins, ethylene, gibberellins, cytokinins, and abscisic acid, have been shown to influence flowering in different species. The critical aspect ofphotoperiodism, however, is the measurement of seasonal time by detecting the lengths of day and night. The discovery of the night break phenomenon, which showed that interruption of the night period with light inhibited flowering in SDPs, established that the length of the dark period is the most critical for initiating a photoperiodic response.

The chemical phytochrome is responsible for measuring the dark period. Phytochrome, found in the leaves of plants, exists in two forms, Pr and Pfr. Prabsorbs red light during the day and is converted to Pfr. Pfr absorbs far-red light during the night and is converted to Pr. The prevailing hypothesis is that Pfr inhibits flowering, and the length of the dark period has to be sufficient for the Pfr to fall below some critical level. When the Pfrfalls below this level, chemical messages are sent to the floral regions, and flowering is initiated.

While phytochrome definitely has been shown to trigger the flowering response, it is not the only chemical involved. It has been shown that a bluelight photoreceptor may also play a role in photoperiodism. In addition, phytochrome is not translocated in the plant. It remains in the leaves. Hence, other chemicals which have not been positively identified are responsible for signalling the photoperiodic response.

PHOTON FLUENCE OF SEED GERMINATION

Experiments were designed to determine the photon fluence requirement of 667 nm light for PhyB-specific germination. WT seeds effectively germinated when irradiated with a fluence higher than 20 mmol m^{-2} of 667 nm light and reached a maximum at a fluence of 200 mmol m^{-2}.

This range of photon fluence is equivalent to the PhyB-specific germination previously reported in A. thaliana ecotype Landsberg erecta. Moreover, seeds of the phyB mutant did not show photo-induction of germination when they were allowed to imbibe water for 3 h, again similar to the phyB mutant of Landsberg erecta.

To determine whether an increase of over-expressed PhyB causes an increase or decrease in the photon fluence required for induction of PhyB- specific germination, we examined the effect of various ranges of photon fluence of R (667 nm) on the photo-induction of RBO and ABO seed germination. The RBO seeds showed similar sensitivity to that of WT seeds. In contrast, ABO seeds showed more sensitive induction of germination than that of WT; photo-induction occurred by irradiation with a fluence higher than 7 mmol m^{-2} of 667 nm light and reached a maximum at a fluence of 70 mmol m^{-2}.

The present results together with data on immunochemical detection of PHYB in seeds of RBO and ABO indicate that the higher the amount of PHYB in seeds, the less photon fluence of R required for the PhyB-specific photo- induction of seed germination.

SHIFT OF EFFECTIVE RANGE OF WAVELENGTH IN ABO

In order to assess whether the over-expression of PHYB leads to a shift in the effective range of wavelength for induction of germination, we examined the effect of various photon fluences of monochromatic light in the spectral range from 300-720 nm on germination using the Okazaki large spectrograph. RBO showed similar sensitivity to WT at all wavelengths examined. Representative fluence-response curves for induction of germination in WT and ABO.

Based on the similarity and differences between the fluence-response curves of WT and of ABO, the range of wavelength was divided into three groups. In the wavelength range 300-350 nm, both WT and ABO showed similar fluence-response curves. In the range 400-690 nm, both reached a maximum of 100 per cent germination, but the ABO seeds germinated with a lower photon fluence than WT seeds. Wavelengths longer than 690 nm showed complicated fluence-response curves. From 694 nm, the final germination percentages of WT seeds decreased, and irradiation with monochromatic light longer than 700 nm did not induce germination within the fluence range examined (less than 100 mmol m^{-2}). In contrast, ABO seeds germinated to 100 per cent by irradiation with monochromatic light from 694-700 nm. From 705-715 nm, the final germination per cent ages of ABO seeds decreased, and irradiation with monochromatic light longer than 720 nm did not induce germination within the fluence range examined (less than 100 mmol m^{-2}, data not shown). Thus, in WT seeds, a critical threshold of response to light wavelength occurred between 700 and 705 nm. In contrast, the corresponding critical wavelength for ABO shifted in the direction of longer wavelength. An explanation for these phenomena may relate to the photoconversion of phytochrome at photostationary state.

Representative fluence-response relationships for WT (s) and ABO (u) seeds are shown.

PHOTOREVERSIBLE INHIBITION OF GERMINATION IN ABO

To determine whether the over-expressed PHYB has any physiological effects on PhyB-specific photoreversible regulation in seeds, we examined R/ FR reversibility of seed germination in WT, ABO and RBO. All of the lines clearly exhibited R/FR reversibility by irradiation with R (5.8 mmol m-2) and then FR (18 mmol m^{-2}).

This result establishes that over expressed PHYB does not interfere with the R/ FR reversible regulation of seed germination within the photon fluence range examined. calculated values of Pfr/Ptot ratio. In theory, the same Pfr/ Ptot ratio given with different light wavelengths should induce the same level of seed germination, if the Pfr/Ptot ratio determines the response. This was demonstrated when the data for different wavelengths were compared within either WT seeds or ABO seeds. However, the germination of ABO seeds was induced at lower Pfr/ Ptot ratio than that of WT seeds. This result strongly suggests that the seed germination is not determined by the Pfr/Ptot ratio itself.

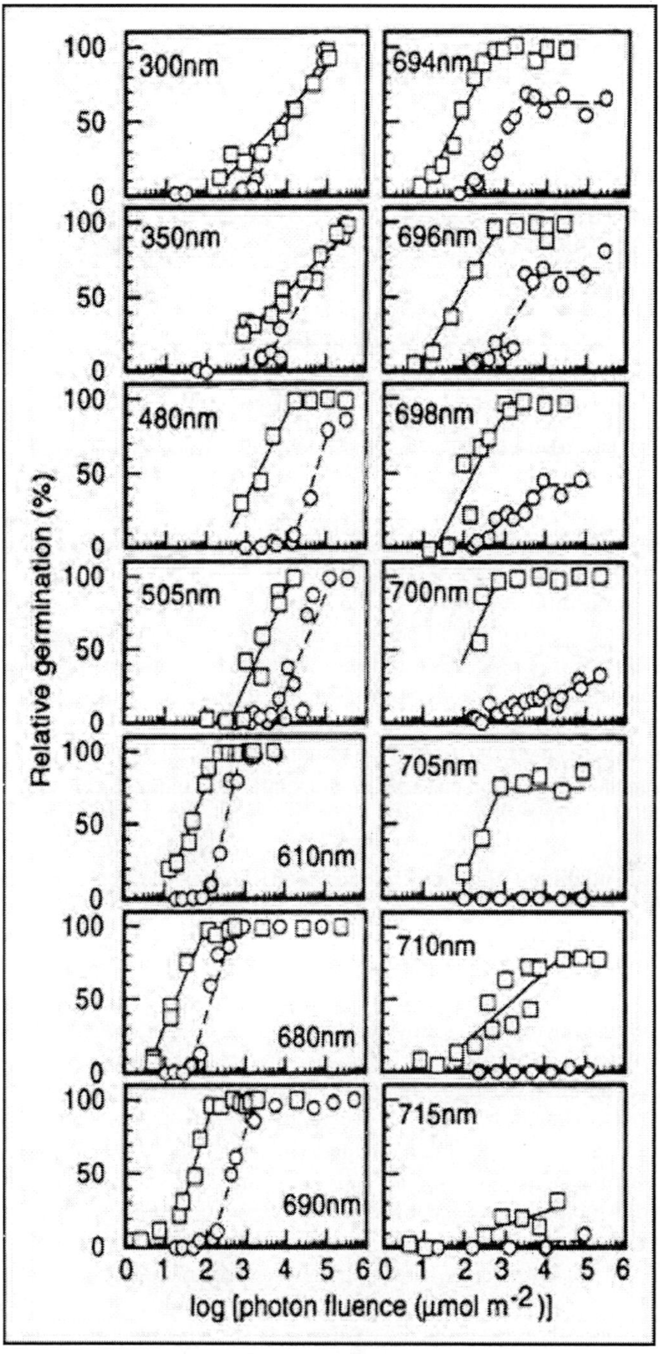

Fig. Effect of Photon Fluence of 300-715 nm light on Induction of PhyB Dependent Seed Germination.

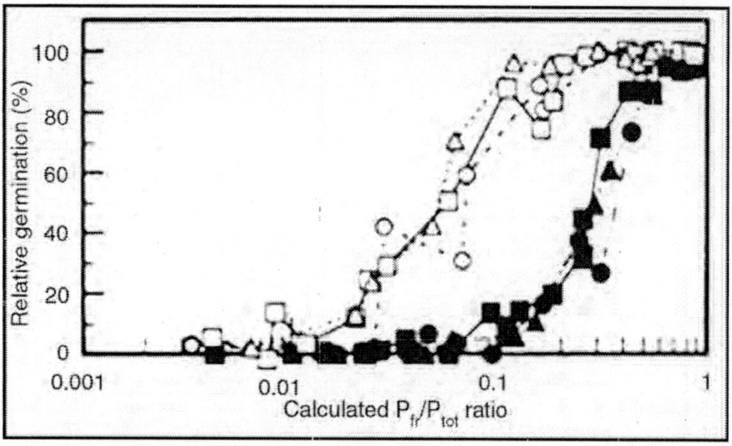

Fig. Relationship between calculated Pfr/Ptot ratios and Germination Percentage for Induction of Germination in WT and ABO seeds.

The germination percentages versus calculated P_{fr}/Ptot ratio. WT seeds irradiated with 505 nm (d), 667 nm (j) and 690 nm (m) and ABO seeds irradiated with 505 nm (s), 667 nm (u), 690 nm (n) are shown.

To analyse the difference in P_{fr}/P_{tot} ratio between WT and ABO at various wavelengths, the photon fluence required for 50 per cent induction of germination were measured and the corresponding P_{fr}/P_{tot} ratio was estimated. Within the wavelength range of 480-698 nm, 50 per cent induction of WT seeds was induced by exposure to a photon fluence which is able to generate a P_{fr}/P_{tot} ratio of 0.21-0.43. In contrast, ABO seeds require a lower fluence which is able to generate P_{fr}/P_{tot} ratio of 0.035-0.056. These calculated values are consistent and confirm again the assertion that seed germination in Arabidopsis is not simply regulated by a mechanism determined exclusively by P_{fr}/P_{tot} ratio.

The shift of the critical threshold of light wavelength for either induction or photoreversible inhibition of germination between WT and ABO seeds is also understood in similar terms. Theoretically, if the P_{fr}/P_{tot} ratio determines seed germination, the critical threshold wavelength for induction of germination in PhyB over-expressing seeds should not change from that of WT seeds.

P_{fr}/P_{tot} ratio is determined by wavelength and photon fluence, irrespective to the absolute amount of phytochrome. However a shift of the critical wavelength in ABO seeds. WT seeds showed a critical wavelength between 700 and 705 nm which should generate a Pfr/Ptot ratio of approximately 0.2 in the photostationary state.

In contrast, ABO seeds showed a longer critical wavelength between 715 and 720 nm which generated a P_{fr}/P_{tot} ratio of approximately 0.04. This difference of critical wavelength again demonstrated that the P_{fr}/P_{tot} ratio did not determine the PhyB-dependent germination of Arabidopsis seed, but that a certain absolute amount of P_{fr} was probably the determining factor.

However, the quantitative relationship between the amount of PhyB and the amount of photon fluence required for PhyB-dependent germination did not show a simple proportional relationship. For example, ABO seeds (which had twice as much PHYB as WT seeds) required less than one-sixth as much R (667 nm) photon fluence for induction of germination and 12 times more FR (750 nm) photon fluence for inhibition of germination relative to WT seeds.

These results suggest that photoperception by the PhyBmonomer molecule is not the only limiting factor for PhyB signal transduction towards seed germination. In the future, measurements of the photochemical parameters of PhyB in vitro, photoconversion and dark reversion kinetics of PhyB in vivo, behaviour of PhyB dimer molecules, and detection of PhyB signal transduction components are expected to provide additional insight into how plants perceive environmental light and adjust seed germination accordingly.

PLANT MATERIALS

Arabidopsis thaliana (L.) Heynh., ecotype Nossen, and two derivative transgenic homozygous lines were used. The cDNA of Arabidopsis PHYB (for ABO) and rice PHYB (for RBO) were fused to the cauliflower mosaic virus 35S promoter. The phyB mutant seeds were obtained by crossing the phyB-1 allele (ecotype Landsberg erecta) into WT (ecotype Nossen) twice.

LIGHT TREATMENT AND GERMINATION ASSAY

Seeds were surface-sterilized and sown onto 0.7 per cent (w/v) aqueous agar medium in plastic Petri dishes in lots of 80-100 seeds each, then immediately exposed to FR (18 mmol m^{-2}), inhibiting PhyB-dependent dark germination as described previously.

They were kept in total darkness for 3-9 h at 23°C before being transferred to the appropriate light treatment. After the light treatment, seeds were kept in darkness for 7 days at 23°C to allow germination to proceed. Germination was scored using a microscope to assess radicle emergence. To normalize experimental differences in germination percentage, the relative germination percentage (R_{li}) was calculated as follows: R_{li} 5 (G_{li} - G_D)/(G_R - G_D), where G_{li} is the germination percentage at each wavelength at each photon fluence, G_D is the germination percentage in darkness, and G_R is the germination percentage upon irradiation with the saturating fluence of R (3 mmol m_{-2}).

Photon fluence was plotted against observed seed germination at 30 different wavelengths from 300 to 820 nm at intervals of 2-50 nm.

Each curve was fitted by the least-squares method. To test photoreversible inhibition of germination, seeds were exposed to saturating R (1.5 mmol m_{-2}) and subsequently irradiated with monochromatic light.

Monochromatic light was generated by the Okazaki large spectrograph. The total fluence was varied by changing the duration (1-30 min) and/or fluence rate (5.0 3 10-4to 3.5 3 101 mmol m_{-2} s-1) of the radiation with threshold boxes and neutral density filters (ND1-50, Hoya, Tokyo, Japan). Fluence rate was measured by a photon density meter or a optical power meter.

❐

6.

Plants and Poor Soil

WHEN a soil is sick, either because its beneficial bacteria do not perform their functions properly, or because of abnormalities in its chemical or physical properties, careful treatment and proper cultural methods may restore it to health. But when a soil becomes sick and unproductive because parasitic forms gain a foothold in it, much greater skill and knowledge are required to cope with the problem. Its solution is of the greatest economic importance to the gardener and to the greenhouse man.

Parasitic fungi, upon finding their way into a soil, do not necessarily interfere with the work of the beneficial bacteria, such as the ammonifiers and nitrifiers, for instance. Nor do they always influence the chemical or physical nature of the soil. Many of them directly attack the crop itself, causing serious diseases in the plants.

DAMPING OFF

This disease is very familiar to every grower of plants. It is peculiar to seedlings or tender plants, and is very prevalent in the greenhouse, the hot bed, the cold frame as well as in the field. It is induced by the presence of definite parasitic fungi, which thrive best in overwatered soils, and when the greenhouse is kept at a comparatively high temperature with poor ventilation. Damping off is also favoured by thick sowing and too much shade in the seed bed. Symptoms of Damping Off. Every experienced grower knows the disease when he sees it. Seedlings freshly damped off are soft and water soaked at the base of the stem. If they are pulled they often break off easily. A more careful examination shows that the root system is entirely decayed, although the upper part of the stem and leaves may still be green, and also possibly fresh.

The degree of prostration in the seedlings is determined by the amount of moisture in the soil. If it is slight, the seedlings will become flabby and wilted before they topple over. With a high moisture content, they are more firm, but become prostrate as soon as infection sets in. The trouble usually begins in pots in the bed, thence spreading in every direction. Damping off is usually caused by several fungi, the chief of which is Pythium de Baryanum Hesse.

The organism was first named and described by Hesse in 1874. Ward found it to be a very prevalent parasite in the garden soils of Europe. In America the fungus was first recognized as of great economic importance by Atkinson. The seedlings of most greenhouse plants may become subject to damping off by Pythium.

When examined under a compound microscope, Pythium de Baryanum is seen to be made up of coarse non-separate, highly granular, irregularly branched hyaline vegetative threads or mycelium. The younger growing threads are more finely granular. The oldest are coarsely granular or more often empty. These threads penetrate the cells of the host, where they obtain its food.

Pythium de Baryanum does not often fruit on the dead seedlings. The fruiting is better observed when the fungus is grown in pure culture. Under normal conditions it produces two forms of spores, conidia and oogonia. The summer spores, or conidia, are swellings formed at the tip of the hyphæ. These swellings readily break off from the mother threads and germinate by sending out a slender tube. This tube penetrates the seedling tissue where it grows and develops and after due incubation reproduces the disease.

The oospore, or sexual spore, is the stage which is most commonly found. The female organ (oogonium) first develops as a terminal enlargement which is cut off by a septum from the mother thread. Next or adjacent to it a slender tube is cut off from the mycelium by a septum. This tube performs the function of the male sexual organ and is known as antheridium. The latter then comes into close contact and empties all its content into the female oogonium. Fertilization thus takes place, and a mature egg, or oospore, or winter resting spore is formed.

The latest investigations have not yet disclosed whether or not Pythium de Baryanum is carried over from year to year by its oospores. It is apparently able to propagate itself indefinitely by its vegetative mycelium.

Of the other fungi which are capable of producing a damping off in the greenhouse or seed bed may be mentioned Sclerotinia libertiana Fckl., Phoma solani Halst., Colletotrichum sp., Fusarium sp., Sclerotium rolfsii Sacc. and Rhizactonia solani Kuhn. Each of these, except the last, will be taken up separately in connection with the study of their respective hosts.

The fungi which cause damping off are introduced into the greenhouse, primarily with sick soil used in the compost, and also with infected manure. The practice of dumping diseased plants and all other infected material in the manure pile cannot be too strongly condemned. Sometimes very lightly infected plants with no visible symptoms of disease in the seed bed may nevertheless act as carriers, and like-wise infect the greenhouse soil.

DAMPING OFF OF CUTTINGS

Greenhouse men are often troubled with a damping off of cuttings. In specific cases this is brought about by parasitic fungi which, however, will be taken up at length under the discussion of the various hosts.

ROOT ROT

Although not so virulent as Pythium, Rhizoctonia is a frequent cause of considerable failure in green-house culture. The fungus causes a damping off of seedlings and cuttings and a serious root rot.

Symptoms. The symptoms of Rhizoctonia rot or wilt do not differ materially from those produced by Pythium de Baryanum. On older plants, however, Rhizoctonia produces cankers or deep lesions which are very characteristic. These are formed on the roots as well as on the base of the stem. The lesions are reddish brown and extend into the cortical or vital layer as well as into the woody tissue.

Fig. Root Rot

There is perhaps no other parasitic fungus which is so widespread and which is capable of attacking such a variety of hosts as Rhizoctonia. The work of Peltier shows that the following greenhouse crops are susceptible to Rhizoctonia: beet, bean, cauliflower, celery, cucumber, egg plant, horseradish, lettuce, muskmelon, pepper, radish, tomato, sweet alyssum, amaranthus, ornamental asparagus, china aster, begonia, candytuft, carnation, coleus, dianthus, lavatera, lobelia, pansy, poinsettia, sweet pea, violet.

Cuttings of the following hosts are also reported by Peltier to damp off from Rhizoctonia: Abutilon hybridum, var. lavitzii, Acalypha wilkesiana, var. bicolour, A. wilkesiana, var. tricolour; A. wilkesiana, var. marginata, Ageratum Piguerua trinervia, Lautolina chamoecyparissus, Sedum spectabile, Althernanthera, Vincia major.

The Organism. In the United States the first extended account of Rhizoctonia was given by Pammel. Many other excellent accounts by American workers have appeared from time to time, to which we shall have occasion to refer later. The genus Rhizoctonia includes several forms of sterile fungi, all of which are distinguished by the manner of growth in pure culture, and by its mycelium. Young hyphae of R. solani Kuhn are at first hyaline, then deepening in colour from a yellowish to a deep brown. The young branches are somewhat narrowed at their point of union with the parent hyphæ and grow in a direction almost parallel with each other. A septum is also laid down at a short distance from the point of union with the parent mycelium.

There is another form of hypha which is made up of barrel shaped cells each of which is capable of germinating like a spore. In pure cultures R. solani produces sclerotia which are first soft, whitish, and which later become hard and dark. The fungus is, carried over from year to year as sclerotia which are able to withstand the effects of heat, cold, drought, or moisture.

Next in importance to Rhizoctonia is a group of fungi which belong to the genus Fusarium. Soils infected with these species of fungi become unfit for tomatoes, sweet peas, etc., thereby causing great financial losses to the greenhouse man. Individual difficulties will be taken up in studying each of these crops separately. As an illustration of a typical

Fusarium sick soil let us consider the wilt of sweet pea. The cause of this trouble is a soil inhabiting fungus, Fusarium lath yri Taub.

Symptoms. The first symptom of the disease is a sudden flagging of the leaves, accompanied by general wilting and collapse of the seedling. Usually upon sowing the seeds a fair percentage germinate and reach the height of about 8 to 10 inches before they are attacked by the fungus. If the collapsed seedlings are allowed to remain on the ground, the stems will soon be covered with the sickle shaped spores. Eventually the decayed tissue rots and is soon invaded by small fruit flies which now begin to distribute the fungus from place to place by carrying its spores.

The Organism. The mycelium of Fusarium lath yri is hyaline, septate and branched. At an early age the mycelial cells round up into countless numbers of chlamydospores. Old cultures are practically one mass of these resting bodies. The spores are of two sorts, the macroconidia which are sickle shaped, 3-4 septate, the microconidia are one celled, minute spherical to elliptical.

ROOT KNOT

Caused by Heterodera radicicola (Greef) Mull. Although root knot is most prevalent in light soils, it may sometimes be found in heavier lands. The trouble is most widespread out of doors in the South-ern States, where the winter is mild. In the North the worm is usually unable to winter over in the open unless it is protected by trash or dead weeds. It is, however, prevalent in greenhouses and is undoubtedly introduced with sick soil brought in from the field.

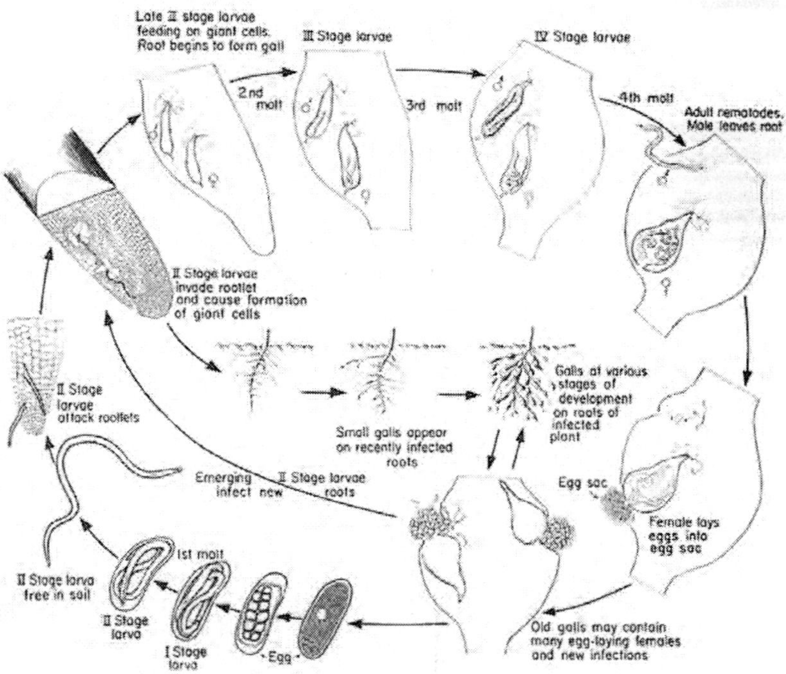

Fig. Root Knot

Symptoms. The disease is characterized by swellings or knots on the roots. These swellings may be variously shaped, and are often mistaken for the true nodules of legumes. Infected plants become stunted, pale, and usually linger for a long time before dying.

The Organism. The nematode is a very minute worm, seldom exceeding one twenty-fifth of an inch in length. It is semi-transparent so that it cannot be easily detected by the naked eye. In searching for the eel worm, it is necessary to break a fresh knot.

Close examination will reveal two types of worms; a spindle shaped worm, the male, and a pearly white pear shaped organism, the female, firmly embedded in the gall tissue. The female is very prolific, depositing no less than 400 to goo eggs during her lifetime.

The eggs are whitish, semi-transparent, bean shaped bodies, and too small to be noticed without the aid of a magnifying glass. The time which elapses until the eggs hatch depends largely upon weather conditions. In warm days the eggs hatch sooner than in cold days.

Upon hatching, the young larvæ either remain in the tissue of the host plant in which they emerge, or, as is more often the case, leave the host and enter the soil. This is the only period during which the worms move about to any great extent in the soil, where they either remain for some length of time or immediately penetrate an-other root of the host.

Maximum, Alyssum odoratum; Coleus, Cuphea phatycentra, Tresine, Petuneia, establishing themselves in the soft cellular structure which is rich in food. The head of the worm is provided with a boring apparatus consisting of a sharply pointed spear, located in the mouth. This structure not only aids it in getting food but is also valuable in helping the young worms to batter through the cell walls before becoming definitely located.

The two sexes during the development are indistinguishable up to fifteen or twenty days, both being spindle shaped. In the molting or shedding of the skin, there is a marked change in the case of the female, especially in the posterior region of the body, which no longer possesses a tail-like appendage. Fertilization occurs soon after this molt, and many radical changes occur in the shape and structure of the organization of the worm.

The fertilized female increases rapidly in breadth and be-comes a pearly white flask or pear-shaped individual. At this stage it is far from being worm- like and may, therefore, be overlooked by one unfamiliar with the life-history of the eel worm. The young male is much like that of the young female larvæ, being spindle shaped in outline.

The male does not cause as much damage to the root tissue as the female, and its purpose in life seems to be only that of fertilizing the female, for after this function has been performed, it is quite probable that the male worm takes no more food.

Omnivorous Nature of the Eel Worm. There are two hundred and thirty- five species of plants known to suffer from the eel worm. This number includes all the important families of the flowering plants. According to Bessey * the following are among the greenhouse plants subject to root knot: bean, beet, cantaloupe, cauliflower, cucumber, egg-plant, lettuce, radish, tomato. For methods of control, see p. 40.

LEAF BLIGHT NEMATODE

Beside the root knot disease which is caused by Heterodera radicicola there is another nematode which confines its injury to foliage only. Of the greenhouse hosts affected by this pest may be mentioned the Begonia, Asplenium nidus-avis, Pteris serrulata avistata, Pteris wimeseth, Pteris tremula, and Pelargonium.

Symptoms. On the Cincinnati begonia the symptoms, according to Clinton, are manifested as numerous small indistinct discolourations limited by the small veinlets. In time, however, the tiny spots enlarge and unite, forming a conspicuous reddish-brown blotch. Frequently infection is manifested as long streaks along the main veins. Often isolated spots occur in the midst of the surrounding healthy tissue.

On Asplenium nidus-avis, the trouble becomes conspicuous in dark brown areas from the base of the leaf near the midrib. These spread up-ward until the entire lower half of the leaf is killed. On Pteris, the spots appear as reddish brown bands reaching out from the midrib to the border, but limited sidewise by the small parallel cross veins.

The Organism. The nematode in question is a slender microscopical worm. The latter chooses the air chamber of the leaf in which to lay its eggs and upon hatching travels around in different parts of the same leaf or to the neighbouring foliage. The worm can travel only when there is a wet film on the leaves.

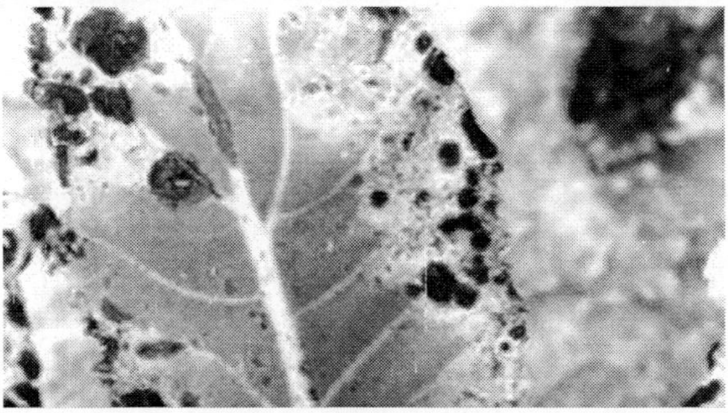

Fig. Leaf Blight Nematode

Control. Immersing fern plants for five minutes in water heated to 122 degrees F. (50 degrees C.) does not seem to injure the ferns, but seems to kill the nematode. All infected leaves should be cut off and burned. Diseased plants should be isolated from healthy ones. Spraying with Bordeaux may also act as a repellant.

SOIL

Soil, the loose material that covers the land surfaces of Earth and supports the growth of plants. In general, soil is an unconsolidated, or loose, combination of inorganic and organic materials. The inorganic components of soil are principally the products of rocks and minerals that have been gradually broken down by weather, chemical action, and other natural processes. The organic materials are composed of debris from plants and from the decomposition of the many tiny life forms that inhabit the soil.

Soils vary widely from place to place. Many factors determine the chemical composition and physical structure of the soil at any given location. The different kinds of rocks, minerals, and other geologic materials from which the soil originally formed play a role. The kinds of plants or other vegetation that grow in the soil are also important. Topography—that is, whether the terrain is steep, flat, or some combination—is another factor. In some cases, human activity such as farming or building has caused disruption. Soils also differ in colour, texture, chemical makeup, and the kinds of plants they can support. Soil actually constitutes a living system, combining with air, water, and sunlight to sustain plant life. The essential process of photosynthesis, in which plants convert sunlight into energy, depends on exchanges that take place within the soil. Plants, in turn, serve as a vital part of the food chain for living things, including humans. Without soil there would be no vegetation— no crops for food, no forests, flowers, or grasslands. To a great extent, life on Earth depends on soil.

The study of different soil types and their properties is called soil science or pedology. Soil science plays a key role in agriculture, helping farmers to select and support the crops on their land and to maintain fertile, healthy ground for planting. Understanding soil is also important in engineering and construction. Soil engineers carry out detailed analysis of the soil prior to building roads, houses, industrial and retail complexes, and other structures.

Soil takes a great deal of time to develop—thousands or even millions of years. As such, it is effectively a nonrenewable resource. Yet even now, in many areas of the world, soil is under siege. Deforestation, over-development, and pollution from humanmade chemicals are just a few of the consequences of human activity and carelessness. As the human population grows, its demand for food from crops increases, making soil conservation crucial.

Quality of Soil

The good or bad qualities of a soil have reference to the needs of the crops which are to be grown upon it, and it is only after a consideration of the requirements of plants that a clear conception can be formed of what characters the soil must possess for it to be a suitable medium on which healthy crops can be raised.

In the first place, soil, to be of any use, must be sufficiently loose and porous to allow the roots of plants to grow and extend freely. It may be so compact that root development is checked or stopped altogether, in which case the plant suffers. On the other hand it should not be too open in texture or the roots do not get a proper hold of the ground and are easily disturbed by wind: moreover such soils are liable to blow away, leaving the underground parts exposed to the air and drought.

The roots like all other parts of plants contain protoplasm or living material, which cannot carry on its functions unless it is supplied with an adequate amount of oxygen: hence the necessity for the continuous circulation of fresh air through the soil.

If the latter is too compact or has its interstices filled with carbon dioxide gas or with water - as is the case when the ground is water-logged - the roots rapidly die of suffocation just as would an animal under the same conditions. There is another point which requires attention. Plants need very considerable amounts of water for their nutrition and growth; the waterholding capacity is, therefore, important. If the soil holds too much it becomes water-logged and its temperature falls below the point for healthy growth, at any rate of the kinds of plants. Usually cultivated on farms and in gardens. If it allows of too free drainage drought sets in and the plants, not getting enough water for their needs, become stunted in size. Too much water is bad, and too little is equally injurious.

In addition, the temperature of the soil largely controls the yield of crops which can be obtained from the land. Soil whose temperature remains low, whether from its northerly aspect or from its high water content or other cause, is unsatisfactory, because the germination of seeds and the general life processes of plants cannot go on satisfactorily except at certain temperatures well above freezing-point.

A good soil should be deep to allow of extensive root development and, in the case of arable soils, easy to work with implements. Even when all the conditions above mentioned in regard to texture, water-holding capacity, aeration and temperature are suitably fulfilled the soil may still be barren: plant foodmaterial is needed. This is usually present in abundance although it may not be available to the plant under certain circumstances, or may need to be replenished or increased by additions to the soil of manures or fertilizers.

CHIEF CONSTITUENTS OF THE SOIL

An examination of the soil shows it to be composed of a vast number of small particles of sand, clay, chalk and humus, in which are generally imbedded larger or smaller stones. It will be useful to consider the nature of the four chief constituents just mentioned and their bearing upon the texture, water- holding capacity and other characters which were referred to in the previous section.

Sand consists of grains of quartz or flint, the individual particles of which are large enough to be seen with the unaided eye or readily felt as gritty grains when rubbed between the finger and thumb. When a little soil is shaken up with water in a tumbler the sand particles rapidly fall to the bottom and form a layer which resembles ordinary sand of the seashore or river banks.

Chemically pure sand is silicon dioxide (SiO_2) or quartz, a clear transparent glass-like mineral, but as ordinarily met with, it is more or less impure and generally coloured reddish or yellowish by oxide of iron.

A soil consisting of sand entirely would be very loose, would have little capacity to retain water, would be liable to become very hot in the daytime and cool at night and would be quite unsuitable for growth of plants.

The term clay is often used by chemists to denote hydrated silicate of alumina ($Al_2O_3.2SiO_2.2H_2O$), of which kaolin or china clay is a fairly pure form. This substance is present in practically all soils but in comparatively small amounts. Even in the soils which farmers speak of as stiff clays it is rarely present to the extent of more than I or 2 per cent.

The word " clay " used in the agricultural sense denotes a sticky intractable material which is found to consist of exceedingly fine particles (generally less than. 005 mm. in diameter) of sand and other minerals derived from the decomposition of rocks, with a small amount of silicate of alumina. The peculiar character which clay possesses is probably due not to its chemical composition but to its physical state. When wet it becomes sticky and almost impossible to move or work with farm implements; neither air nor water can penetrate freely. In a dry state it becomes hard and bakes to a brick. It holds water well and is consequently cold, needing the application of much heat to raise its temperature. It is obvious, therefore, that soil composed entirely of clay is as useless as pure sand so far as the growth of crops upon it is concerned.

Chalk consists, when quite pure, of calcium carbonate (CaC03), a white solid substance useful in small amounts as a plant foodmaterial, though in excess detrimental to growth. Alone, even when broken up into small pieces, it is unsuitable for the growth of plants.

Humus, the remaining constituent of soil, is the term used for the decaying vegetable and animal matter in the soil. A good illustration of it is peat. Its water-holding capacity is great, but it is often acid, and when dr y it is light and incapable of supporting the roots of plants properly. Few of the commonly cultivated crops can live in a soil consisting mainly of humus.

From the above account it will be understood that not one of the four chief soil constituents is in itself of value for the growth of crops, yet when they are mixed, as they usually are in the soils met with in nature, one corrects the deficiencies of the other. A perfect soil would be such a blend of sand, clay, chalk and humus as would contain sufficient clay and humus to prevent drought, enough sand to render it pervious to fresh air and prevent waterlogging, chalk enough to correct the tendency to acidity of the humus present, and would have within it various substances which would serve as food-materials to the crops.

COMPOSITION OF SOILS

Soils comprise a mixture of inorganic and organic components: minerals, air, water, and plant and animal material. Mineral and organic particles generally compose roughly 50 percent of a soil's volume. The other 50 percent consists of pores—open areas of various shapes and sizes.

Networks of pores hold water within the soil and also provide a means of water transport. Oxygen and other gases move through pore spaces in soil. Pores also serve as passageways for small animals and provide room for the growth of plant roots.

Inorganic Material

The mineral component of soil is made up of an arrangement of particles that are less than 2.0 mm (0.08in) in diameter. Soil scientists divide soil particles, also known as soil separates, into three main size groups: sand, silt, and clay.

According to the classification scheme used by the United States Department of Agriculture (USDA), the size designations are: sand, 0.05 to 2.00 mm (0.002 to 0.08 in); silt 0.002 to 0.05 mm (0.00008 to 0.002 in); and clay, less than 0.002 mm (0.00008 in). Depending upon the rock materials from which they were derived, these assorted mineral particles ultimately release the chemicals on which plants depend for survival, such as potassium, calcium, magnesium, phosphorus, sulfur, iron, and manganese.

Organic Material

Organic materials constitute another essential component of soils. Some of this material comes from the residue of plants, the remains of plant roots deep within the soil, or materials that fall on the ground, such as leaves on a forest floor. These materials become part of a cycle of decomposition and decay, a cycle that provides important nutrients to the soil. In general, soil fertility depends on a high content of organic materials. Even a small area of soil holds a universe of living things, ranging in size from the fairly large to the microscopic: earthworms, mites, millipedes, centipedes, grubs, termites, lice, springtails, and more. And even a gram of soil might contain as many as a billion microbes—bacteria and fungi too small to be seen with the naked eye.

All these living things form a complex chain: Larger creatures eat organic debris and excrete waste into the soil, predators consume living prey, and microbes feed on the bodies of dead animals. Bacteria and fungi, in particular, digest the complex organic compounds that make up living matter and reduce them to simpler compounds that plants can use for food. A typical example of bacterial action is the formation of ammonia from animal and vegetable proteins. Other bacteria oxidize the ammonia to form nitrogen compounds called nitrites, and still other bacteria act on the nitrites to form nitrates, another type of nitrogen compound that can be used by plants. Some types of bacteria are able to fix, or extract, nitrogen directly from the air and make it available in the soil.Ultimately, the decay of plant and animal material results in the formation of a dark-coloured organic matter known as humus. Humus, unlike plant residues, is generally resistant to further decomposition.

CHEMICAL COMPOSITION OF THE SOIL

It has been found by experiment that plants need for their nutritive process and their growth, certain chemical elements, namely, carbon, hydrogen, oxygen, nitrogen, sulphur, phosphorus, potassium, magnesium, calcium and iron. With the exception of the carbon and a small proportion of the oxygen and nitrogen, which may be partially derived from the air, these elements are taken from the soil by crops.

The following table shows the amounts of the chief constituents removed by certain crops in lb per acre: - Plants also remove from the soil silicon, sodium, chlorine, and other elements which are, nevertheless, found to be unessential for the growth and may therefore be neglected here.

Leguminous crops take some of the nitrogen which they require from the air, but most plants obtain it from the nitrates present in the soil. The sulphur exists in the soil chiefly in the form of sulphates of magnesium, calcium and other metals; the phosphorus mainly as phosphates of calcium, magnesium and iron; the potash, soda and other bases as silicates and nitrates; calcium and magnesium carbonates are also common constituents of many soils. In the ordinary chemical analyses of the soil determinations are made of the nitrogen and various carbonates present as well as of the amount of phosphoric acid, potash, soda, magnesia and other components soluble in strong hydrochloric acid.

Soil Formation

Soil formation is an ongoing process that proceeds through the combined effects of five soil-forming factors: parent material, climate, living organisms, topography, and time. Each combination of the five factors produces a unique type of soil that can be identified by its characteristic layers, called horizons. Soil formation is also known as pedogenesis (from the Greek words pedon, for “ground,” and genesis, meaning “birth” or “origin”).

Parent Material

The first step in pedogenesis is the formation of parent material from which the soil itself forms. Roughly 99 percent of the world’s soils derive from mineral-based parent materials that are the result of weathering, the physical disintegration and chemical decomposition of exposed bedrock. The small percentage of remaining soils derives from organic parent materials, which are the product of environments where organic matter accumulates faster than it decomposes. This accumulation can occur in marshes, bogs, and wetlands.

Bedrock itself does not directly give rise to soil. Rather, the gradual weathering of bedrock, through physical and chemical processes, produces a layer of rock debris called regolith.

Further weathering of this debris, leading to increasingly smaller and finer particles, ultimately results in the creation of soil.In some instances, the weathering of bedrock creates parent materials that remain in one place. In other cases, rock materials are transported far from their source—blown by wind, carried by moving water, and borne inside glaciers.

Climate

Climate directly affects soil formation. Water, ice, wind, heat, and cold cause physical weathering by loosening and breaking up rocks. Water in rock crevices expands when it freezes, causing the rocks to crack.

Rocks are worn down by water and wind and ground to bits by the slow movement of glaciers. Climate also determines the speed at which parent materials undergo chemical weathering, a process in which existing minerals are broken down into new mineral components. Chemical weathering is fastest in hot, moist climates and slowest in cold, dry climates.

Climate also influences the developing soil by determining the types of plant growth that occur. Low rainfall or recurring drought often discourage the growth of trees but allow the growth of grass. Soils that develop in cool rainy areas suited to pines and other needle-leaf trees are low in humus.

Living Organisms

As the parent material accumulates, living things gradually gain a foothold in it. The arrival of living organisms marks the beginning of the formation of true soil. Mosses, lichens, and lower plant forms appear first. As they die, their remains add to the developing soil until a thin layer of humus is built up. Animals' waste materials add nutrients that are used by plants. Higher forms of plants are eventually able to establish themselves as more and more humus accumulates. The presence of humus in the upper layers of a soil is important because humus contains large amounts of the elements needed by plants.

Living organisms also contribute to the development of soils in other ways. Plants build soils by catching dust from volcanoes and deserts, and plants' growing roots break up rocks and stir the developing soil. Animals also mix soils by tunneling in them.

WATER IN THE SOIL

The importance of an adequate supply of water to growing crops cannot well be over-estimated. During the life of a plant there is a continuous stream of water passing through it which enters by the root-hairs in the soil and after passing along the stem is given off from the 'stomata of the leaves into the open air above ground. It has been estimated that an acre of cabbage will absorb from the land and transpire from its leaves more than ten tons of water per day when the weather is fine.

In addition to its usefulness in maintaining a turgid state of the young cells without which growth cannot proceed, water is itself a plant food-material and as absorbed from the soil contains dissolved in it all the mineral food constituents needed by plants for healthy nutrition. Without a sufficient supply plants remain stunted and the crop yield is seriously reduced, as we see in dry seasons when the rainfall is much below the average.

If one condition is more necessary than another for good crops it is a suitable supply of water, for no amount of manuring or other treatment of the soil will make up for a deficient rainfall. The amount needed for the most satisfactory nutrition varies with different plants. In the case of fair average farm crops it has been shown that for the production of one ton of dry matter contained in them from 300 to 500 tons of water has been absorbed and utilized by the plants. This may be more than the rainfall, in which case irrigation or special control of the water supply may be necessary.

The water-holding capacity of a soil depends upon the amount of free space between the particles of which it is composed into which water can enter. In most cases this amounts to from 30 to 50 per cent of the volume of the soil.

When the pore-space of the soil is filled with water it becomes water- logged and few plants can effect absorption by their roots under such conditions. The root-hairs die from want of air, and the whole plant soon suffers. Fields of wheat and other cereals rarely recover after a week's submergence, but orchards and many trees when at rest in winter withstand a flooded or water-logged condition of the soil for two or three weeks without damage.

The most satisfactory growth is maintained when the amount of water present is not more than 40 to 60 per cent of what would saturate it. Under such conditions each particle of soil is surrounded by a thin film of water and in the pore-space air can freely circulate. It is from such films that the root- hairs absorb all that plants require for their growth.

The movement of water into the root-hairs is brought about by the osmotic action of certain salts in their cell-sap. Crops are, however, unable to absorb all the water present in the soil, for when the films become very thin they are held more firmly or cling with more force to the soil particles and resist the osmotic action of the root-hairs. Plants have been found to wither and die in sandy soils containing i a per cent of water, and in clay soils in which there was still present 8 per cent of water.

When a long glass tube open at both ends is filled with soil and one end is dipped in a shallow basin of water, the water is found to move upwards through the soil column just as oil will rise in an ordinary lamp wick. By this capillary action water may be transferred to the upper layers of the soil from a depth of several feet below the surface. In this manner plants whose roots descend but a little way in the ground are enabled to draw on deep supplies.

Not only does water move upwards, but it is transferred by capillarity in all directions through the soil. The amount and speed of movement of water by this means, and the distance to which it may be carried, depend largely upon the fineness of the particles composing the soil and the spaces left between each. The ascent of water is most rapid through coarse sands, but the height to which it will rise is comparatively small. In clays whose particles are exceedingly minute the water travels very slowly but may ultimately reach a height of many feet above the level of the " water-table " below.

While this capillary movement of water is of great importance in supplying the needs of plants it has its disadvantages, since water may be transferred to the surface of the soil, where it evapo rates into the air and is lost to the land or the crop growing upon it..

The loss in this manner was found to be in one instance over a pound. of water per day per square foot of surface, the " water-table " being about 4 or 5 ft. below.

One of the most effective means of conserving soil moisture is by " mulching," i.e. by covering the surface of the soil with some loosely compacted material such as straw, leaf-refuse or stablemanure. The space between the parts of such substances is too large to admit of capillary action; hence the water conveyed to the surface of the soil is prevented from passing upwards any further except by slow evaporation through the mulching layer.

A loose layer of earth spread over the surface of the soil acts in the same way, and a similarly effective mulch may be prepared by hoeing the soil, or stirring it to a depth of one or two inches with harrows or other implements. The hoe and harrow are therefore excellent tools for use in dry weather. Rolling the land is beneficial to young crops. in dry weather, since it promotes capillary action by reducing the soil spaces. It should, however, be followed by a light hoeing or harrowing.

In the semi-arid regions of the United States, Argentina and other countries where the average annual rainfall lies between ioa to 20 in., irrigation is necessary to obtain full crops every year. Good crops, however, can often be grown in such areas without irrigation if attention is paid to the proper circulation of water in the soil and means for retaining it or preventing excessive loss by evaporation.

Of course care must be exercised in the selection of plants - such as sorghum, maize, wheat, and alfalfa or lucerne - which are adapted to dry conditions and a warm climate.

So far as the water-supply is concerned - and this is what ultimately determines the yield of crops - the rain which falls upon the soil should be made to enter it and percolate rapidly through its interstices. A deep porous bed in the upper layers is essential, and this should consist of fine particles which lie close to each other without any tendency to stick together and " puddle " after heavy showers. Every effort should be made to prepare a good mealy tilth by suitable ploughing, harrowing and consolidation.

In the operation of ploughing the furrow slice is separated from the soil below, and although in humid soils this layer may be left to settle by degrees, in semi-arid regions this loosened layer becomes. dry if left alone even for a few hours and valuable water evaporates. into the air. To prevent this various implements, such as disk harrows and specially constructed rollers, may be used to consolidate the upper stirred portion of the soil and place it in close capillary relationship with the lower unmoved layer.

If the soil is allowed to become dry and pulverized, rain is likely to run off or " puddle the surface without penetrating it more than a very short distance. Constant hoeing or harrowing to maintain a natural soil mulch layer of 2 or 3 in. deep greatly conserves the soil water below. In certain districts where the rainfall is low a crop can only be obtained once every alternate year, the intervening season being devoted to tillage with a view of getting the rain into the soil and retaining it there for the crop in the following year.

Soil scientists also characterize soils according to how effectively they retain and transport water. Once water enters the soil from rain or irrigation, gravity comes into play, causing water to trickle downward. Water is also taken up in great quantities by the roots of plants: Plants use anywhere from 200 to 1,000 kg (440 to 2,200 lb) of water in the formation of 1 kg (2.2 lb) of dry matter. Soils differ in their capacity to retain moisture against the pull exerted by gravity and by plant roots. Coarse soils, such as those consisting of mostly of sand, tend to hold less water than do soils with finer textures, such as those with a greater proportion of clays.

Water also moves through soil pores independently of gravity. This movement can occur via capillary action, in which water molecules move because they are more attracted to the pore walls than to one another. Such movement tends to occur from wetter to drier areas of the soil. The movement from soil to plant roots can also depend on how tightly water molecules are bound to soil particles.

The attraction of water molecules to each other is an example of cohesion. The attraction of water molecules to other materials, such as soil or plant roots, is a type of adhesion. These effects, which determine the so-called matric potential of the soil, depend largely on the size and arrangement of the soil particles. Another factor that can affect water movement is referred to as the osmotic potential. The osmotic potential hinges on the amount of dissolved salts in the soil. Soils high in soluble salt tend to reduce uptake of water by plant roots and seeds. The sum of the matric and osmotic potentials is called the total water potential. In soil, water carries out the essential function of bringing mineral nutrients to plants. But the balance between water and air in the soil can be delicate. An overabundance of water will saturate the soil and fill pore spaces needed for the transport of oxygen. The resulting oxygen deficiency can kill plants. Fertile soils permit an exchange between plants and the atmosphere, as oxygen diffuses into the soil and is used by roots for respiration. In turn, the resulting carbon dioxide diffuses through pore spaces and returns to the atmosphere.

This exchange is most efficient in soils with a high degree of porosity. For farmers, gardeners, landscapers, and others with a professional interest in soil health, the process of aeration—making holes in the soil surface to permit the exchange of air—is a crucial activity. The burrowing of earthworms and other soil inhabitants provides a natural and beneficial form of aeration.

Bacteria in the Soil

Recent science has made much progress in the investigation of the micro- organisms of the soil. Whereas the soil used to be looked upon solely as a dead, inert material containing certain chemical substances which serve as food constituents. of the crops grown upon it, it is now known to be a place of habitation for myriads of minute living organisms upon whose activity much of its fertility depends.

They are responsible for many important chemical processes which make the soil constituents more available and better adapted to the nutrition of crops. One cubic centimetre of soil taken within a foot or so from the surface contains from II to 2 millions of bacteria of many different kinds, as well as large numbers of fungi. In the lower depths of the soil the numbers. decrease, few being met with at a depth of 5 or 6 ft.

The efficiency of many substances, such as farm-yard manure,. guanos, bone-meal and all other organic materials, which are spread over or dug or ploughed into the land for the benefit of farm and' garden crops, is bound up with the action of these minute living beings.

Without their aid most manures would be useless for plant growth. Farm-yard manure, guanos and other fertilizers. undergo decomposition in the soil and become broken down into compounds of simple chemical composition better suited for absorption by the roots of crops, the changes involved being directly due to the activity of bacteria and fungi. Much of the work carried on by these organisms is not clearly understood; there are, however, certain processes which have been extensively investigated and to these it is necessary to refer.

It has been found by experiment that the nitrogen needed by practically all farm crops except leguminous ones is best supplied in the form of a nitrate; the rapid effect of nitrate of soda when used' as a top dressing to wheat or other plants is well known to farmers.

It has long been known that when organic materials such as the dung and urine of animals, or even the bodies of animals and plants, are applied to the soil, the nitrogen within them becomes oxidized, and ultimately appears in the form of nitrate of lime, potash or some other base. The nitrogen in decaying roots, in the dead stems. and leaves of plants, and in humus generally is sooner or later changed into a nitrate, the change being effected by bacteria. That: the action of living organisms is the cause of the production of nitrates is supported by the fact that the change does not occur when the soil is heated nor when it is treated with disinfectants which destroy or check the growth and life of bacteria.

The process resulting in the formation of nitrates in the soil is spoken of as nitrification. The steps in the breaking down of the highly complex nitrogenous proteid compounds contained in the humus of the soil, or applied to the latter by the farmer in the form of dung and organic refuse generally, are many and varied; most frequently the insoluble proteids are changed by various kinds of putrefactive bacteria into soluble proteids (peptones, &c.), these into simpler amido-bodies, and these again sooner or later into compounds 7of ammonia.

The urea in urine is also rapidly converted by the uro-bacteria into ammonium carbonate. The compounds of ammonia thus formed from the complex substances by many varied kinds of micro-organisms are ultimately oxidized into nitrates.

The change takes place in two stages and is effected by two special groups of nitrifying bacteria, which are present in all soils. In the first stage the ammonium compounds are oxidized to nitrites by the agency of very minute motile bacteria belonging to the genus Nitrosomonas. The further oxidation of the nitrite to a nitrate is effected by bacteria belonging to the genus Nitrobacter.

Several conditions must be fulfilled before nitrification can occur. In the first place an adequate temperature is essential; at 5° or 6° C. (40-43° F.) the process is stopped, so that it does not go on in winter. In summer, when the temperature is about 24° C. (75° F.), nitrification proceeds at a rapid rate.

The organisms do not carry on their work in soils deficient in air; hence the process is checked in water-logged soils. The presence of a base such as lime or magnesia (or their carbonates) is also essential, as well as an adequate degree of moisture: in dry soils nitrification ceases.

It is the business of the farmer and gardener to promote the activity of these organisms by good tillage, careful drainage and occasional application of lime to soils which are deficient in this substance. It is only when these conditions are attended to that decay and nitrification of dung, guano, fish- meal, sulphate of ammonia and other manures take place, and the constituents which they contain become available to the crops for whose benefit they have been applied to the land.

Nitrates are very soluble in water and are therefore liable to be washed out of the soil by heavy rain. They are, however, very readily absorbed by growing plants, so that in summer, when nitrification is most active, the nitrates produced are usually made use of by crops before loss by drainage takes place. In winter, however, and in fallows loss takes place in the subsoil water.

There is also another possible source of loss of nitrates through the activity of denitrifying bacteria. These organisms reduce nitrates to nitrites and finally to ammonia and gaseous free nitrogen which escapes into the atmosphere. Many bacteria are known which are capable of denitrification, some of them being abundant in fresh dung and upon old straw. They can, however, only carry on their work extensively under anaerobic conditions, as in waterlogged soils or in those which are badly tilled, so that there is but little loss of nitrates through their agency.

An important group of soil organisms are now known which have the power of using the free nitrogen of the atmosphere for the formation of the complex nitrogenous compounds of which their bodies are largely composed. By their continued action the soil becomes enriched with nitrogenous material which eventually through the nitrification process becomes available to ordinary green crops. This power of " fixing nitrogen," as it is termed, is apparently not possessed by higher green plants.

The bacterium, Clostridium pasteurianum, common in most soils, is able to utilize free nitrogen under anaerobic conditions, and an organism known as Azotobacter chroococcum and some others closely allied to it, have similar powers which they can exercise under aerobic conditions. For the carrying on of their functions they all need to be supplied with carbohydrates or other carbon compounds which they obtain ordinarily from humus and plant residues in the soil, or possibly in some instances from carbohydrates manufactured by minute green algae with which they live in close union.

Certain bacteria of the nitrogenfixing class enter into association with the roots of green plants, the best-known examples being those which are met with in the nodules upon the roots of clover, peas, beans, sainfoin and other plants belonging to the leguminous order.

That the fertility of land used for the growth of wheat is improved by growing upon it a crop of beans or clover has been long recognized by farmers. The knowledge of the cause, however, is due to modern investigations. When wheat, barley, turnips and similar plants are grown, the soil upon which they are cultivated becomes depleted of its nitrogen; yet after a crop of clover or other leguminous plants the soil is found to be richer in nitrogen than it was before the crop was grown. This is due to the nitrogenous root residues left in the land. Upon the roots of leguminous plants characteristic swollen nodules or tubercles are present.

These are found to contain large numbers of a bacterium termed Bacillus radicicola or Pseudomonas radicicola. The bacteria, which are present in almost all soils, enter the root-hairs of their host plants and ultimately stimulate the production of an excrescent nodule, in which they live. For a time after entry they multiply, obtaining the nitrogen necessary for their nutrition and growth from the free nitrogen of the air, the carbohydrate required being supplied by the pea or clover plant in whose tissues they make a home.

The nodules increase in size, and analysis shows that they are exceedingly rich in nitrogen up to the time of flowering of the host plant. During this period the bacteria multiply and most of them assume a peculiar thickened or branched form, in which state they are spoken of as bacteroids.

Later the nitrogen-content of the nodule decreases, most of the organisms, which are largely composed of proteid material, becoming digested and transformed into soluble nitrogenous compounds which are conducted to the developing roots and seeds. After the decay of the roots some of the unchanged bacteria are left in the soil, where they remain ready to infect a new leguminous crop.

The nitrogen-fixing nodule bacteria can be cultivated on artificial media, and many attempts have been made to utilize them for practical purposes. Pure cultures may be made and after dilution in water or other liquid can be mixed with soil to be ultimately spread over the land which is to be infected.

The method of using them most frequently adopted consists in applying them to the seeds of leguminous plants before sowing, the seed being dipped for a time in a liquid containing the bacteria. In this manner organisms obtained from red clover can be grown and applied to the seed of red clover; and similar inoculation can be arranged for other species, so that an application of the bacteria most suited to the particular crop to be cultivated can be assured.

In many cases it has been found that inoculation, whether of the soil or of the seed, has not made any appreciable difference to the growth of the crop, a result no doubt due to the fact that the soil had already contained within it an abundant supply of suitable organisms.

But in other instances greatly increased yields have been obtained where inoculation has been practised. More or less pure cultures of the nitrogen- fixing bacteria belonging to the Azotobacter group have been tried and recommended for application to poor land in order to provide a cheap supply of nitrogen.

The application of pure cultures of bacteria for improving the fertility of the land is still in an experimental stage. There is little doubt, however, that in the near future means will be devised to obtain the most efficient work from these minute organisms, either by special artificial cultivation and subsequent application to the soil, or by improved methods of encouraging their healthy growth and activity in the land where they already exist.

Improvement of Soils

The fertility of a soil is dependent upon a number of factors, some of which, such as the addition of fertilizers or manures, increase the stock of available food materials in the soil while others, such as application of clay or humus, chiefly influence the fertility of the land by improving its physical texture.

The chief processes for the improvement of soils are:liming, claying and marling, warping, paring and burning, and green manuring. Most of these more or less directly improve the land by adding to it certain plant food constituents which are lacking, but the effect of each process is in reality very complex. In the majority of cases the good results obtained are more particularly due to the setting free of " dormant " or " latent " food constituents and to the amelioration of the texture of the soil, so that its aeration, drainage, temperature and water-holding capacity are altered for the better.

The material which chemists call calcium carbonate is met with in a comparatively pure state in chalk. It is present in variable amounts in limestones of all kinds, although its white ness may there be masked by the presence of iron oxide and other coloured substances. Carbonate of lime is also a constituent to a greater or lesser extent in almost all soils. In certain sandy soils and in a few stiff clays it may amount to less than 4 per cent, while in others in limestone and chalk districts there may be 50 to 80 per cent present. Pure carbonate of lime when heated loses 44 per cent of its weight, the decrease being due to the loss of carbon dioxide gas.

The resulting white product is termed calcium oxide lime, burnt lime, quicklime, cob lime, or caustic lime. This substance absorbs and combines with water very greedily, at the same time becoming very hot, and falling into a fine dry powder,' calcium hydroxide or slaked lime, which when left in the open slowly combines with the carbon dioxide of the air and becomes calcium carbonate, from which we began.

When recommendations are made about liming land it is necessary to indicate more precisely than is usually done which of the three classes of material named above - chalk, quicklime or slaked lime - is intended. Generally speaking the oxide or quicklime has a more rapid and greater effect in modifying the soil than slaked lime, and this again greater than the carbonate or chalk.

Lime in whatever form it is applied has a many-sided influence in the fertility of the land. It tends to improve the tilth and the capillarity of the soil by binding sands together somewhat and by opening up clays. If applied in too great an amount to light soils and peat land it may do much damage by rendering them too loose and open. The addition of small quantities of lime, especially in a caustic form, to stiff greasy clays makes them much more porous and pliable. A lump of clay, which if dried would become hard and intractable, crumbles into pieces when dried after adding to it 2 per cent of lime. The lime causes the minute separate particles of clay to flocculate or group themselves together into larger compound grains between which air and water can percolate more freely.

It is this power of creating a more crumbly tilth on stiff clays that makes lime so valuable to the farmer. Lime also assists in the decomposition of the organic matter or humus in the soil and promotes nitrification; hence it is of great value after green manuring or where the land contains much humus from the addition of bulky manures such as farm-yard dung.

This tendency to destroy organic matter makes the repeated application of lime a pernicious practice, especially on land which contains little humus to begin with. The more or less dormant nitrogen and other constituents of the humus are made immediately available to the succeeding crop, but the capital of the soil is rapidly reduced, and unless the loss is replaced by the addition of more manures the land may become sterile. Although good crops may follow the application of lime, the latter is not a direct fertilizer or manure and is no substitute for such. Its best use is obtained on land in good condition, but not where the soil is pcor.

When used on light dry land it tends to make the land drier, since it destroys the humus which so largely assists in keeping water in the soil. Lime is a base and neutralizes the acid materials present in badly drained meadows and boggy pastures. Weeds, therefore, which need sour conditions for development are checked by liming and the better grasses and clovers are encouraged. It also sets free potash and possibly other useful plant food-constituents of the soil. Liming tends to produce earlier crops and destroys the fungus which causes finger-and-toe or club-root among turnips and cabbages.

Land which contains less than about z /c, of lime usually needs the addition of this material. The particular form in which lime should be applied for the best results depends upon the nature of the soil. In practice the proximity to chalk pits or lime kilns, the cost of the lime and cartage, will determine which is most economical.

Generally speaking light poor lands deficient in organic matter will need the less caustic form or chalk, while quicklime will be most satisfactory on the stiff clays and richer soils. On the stiff soils overl y ing the chalk it was formerly the custom to dig pits through the soil to the rock below. Shafts 20 or 30 ft. deep were then sunk, and the chalk taken from horizontal tunnels was brought to the surface and spread on the land at the rate of about 60 loads per acre. Chalk should be applied in autumn, so that it may be split by the action of frost during the winter.

Quicklime is best applied, perhaps, in spring at the rate of one ton per acre every six or eight years, or in larger doses-4 to 8 tons - every 15 to 20 years. Small dressings applied at short intervals give the most satisfactory results. The quicklime should be placed in small heaps and covered with soil if possible until it is slacked and the lumps have fallen into powder, after which it may be spread and harrowed in. Experiments have shown that excellent effects can be obtained by applying 5 or 6 cwt. of ground quicklime.

Gas-lime is a product obtained from gasworks where quicklime is used to purify the gas from sulphur compounds and other objectionable materials. It contains a certain amount of unaltered caustic lime and slacked lime, along with sulphates and sulphides of lime, some of which have an evil odour. As some of these sulphur compounds have a poisonous effect on plants, gas- lime cannot be applied to land directly without great risk or rendering it incapable of growing crops of any sort - even weeds - for some time.

It should therefore be kept a year or more in heaps in some waste corner and turned over once or twice so that the air can gain access to it and oxidize the poisonous ingredients in it. Many soils of a light sandy or gravelly or peaty nature and liable to drought and looseness of texture can be improved by the addition of large amounts of clay of an ordinary character.

Similarly soils can be improved by applying to them marl, a substance consisting of a mixture of clay with variable proportions of lime. Some of the chalk marls, which are usually of a yellowish or dirty grey colour, contain clay and 50 to 80 per cent of carbonate of lime with a certain proportion of phosphate of lime. Such a material would not only have an influence on the texture of the land but the lime would reduce the sourness of the land and the phosphate of lime supply one of the most valuable of plant foodconstituents.

The beneficial effects of marls may also be partially due to the presence in them of available potash. Typical clay-marls are tenacious, soapy clays of yellowish-red or brownish colour and generally contain less than 50 per cent of lime. When dry they crumble into small pieces which can be readily mixed with the soil by ploughing.

Many other kinds of marls are described; some are of a sandy nature, others stony or full of the remains of small shells. The amount and nature of the clay or marl to be added to the soil will depend largely upon the original composition of the latter, the lighter sands and gravel requiring more clay than those of firmer texture. Even stiff soils deficient in lime are greatly improved in fertility by the addition of marls.

In some cases as little as 40 loads per acre have been used with benefit, in others 180 loads have not been too much. The material is dug from neighbouring pits or sometimes from the fields which are to be improved, and applied in autumn and winter. When dry and in a crumbly state it is harrowed and spread and finally ploughed in and mixed with the soil.

On some of the strongest land it was formerly the practice to add to and plough into it burnt clay, with the object of making the land work more easily.

The burnt clay moreover carried Cl ay with it potash and other materials in a state readily available to the crops. The clay is dug from the land or from ditches or pits and placed in heaps of 60 to ioo loads each, with faggot wood, refuse coals or other fuel.

Great care is necessary to prevent the heaps from becoming too hot, in which case the clay becomes baked into hard lumps of brick-like material which cannot be broken up.

With careful management, however, the clay dries and bakes, becoming slowly converted into lumps which readily crumble into a fine powder, in which state it is spread over and worked into the land at the rate of 40 loads per acre.

The paring and burning of land, although formerly practised as an ordinary means of improving the texture and fertility of arable fields, can now only be looked upon as a practice p to be adopted for the purpose of bringing rapidly into cultivation very foul leys or, land covered with a coarse turf. The practice is confined to poorer types of land, such as heaths covered with furze and bracken or fens and clay areas smothered with rank grasses and sedges.

To reduce such land to a fit state for the growth of arable crops is very difficult and slow without resort to paring and burning. The operation consists of paring off the tough sward to a depth of I to 2 in. just sufficient to effectually damage the roots of the plants forming the sward and then, after drying the sods and burning them, spreading the charred material and ashes over the land.

The turf is taken off either with the breast plough - a paring tool pushed forward from the breast or thighs by the workman - or with specially constructed paring ploughs or shims. The depth of the sod removed should not be too thick or burning is difficult and too much humus is destroyed unnecessarily, nor should it be too thin or the roots of the herbage are not effectually destroyed.

The operation is best carried out in spring and summer. After being pared off the turf is allowed to dry for a fortnight or so and is then placed in small heaps a yard or two wide at the base, a little straw or wood being put in the middle of each heap, which is then lighted.

As burning proceeds more turf is added to the outside of the heaps in such a manner as to allow little access of air. Every care should be taken to burn and char the sod thoroughly without permitting the heap to blaze. The ashes should be spread as soon as possible and covered by a shallow ploughing. The land is then usually sown with some rapidly growing green crop, such as rape, or with turnips. Paring and burning improves the texture of clay lands, particularly if draining is carried out at the same time. It tends to destroy insects and weeds, and gets rid of acidity of the soil. No operation brings old turf into cultivation so rapidly. Moreover the beneficial effects are seen in the first crop and last for many years.

Many of the mineral plant food-constituents locked up in the coarse herbage and in the upper layers of the soil are made immediately available to crops. The chief disadvantage is the loss of nitrogen which it entails, this element being given off into the air in a free gaseous state.

It is best adapted for application to clays and fen lands and should not be practised on shallow light sands or gravelly soils, since the humus so necessary for the fertility of such areas is reduced too much and the soil rendered too porous and liable to suffer from drought.

Many thousands of acres of low-lying peaty and sandy land adjoining the tidal rivers which flow into the Humber have been improved by a process termed " warping." The warp consists of fine muddy sediment which is suspended in the tidal river water and appears to be derived from material scoured from the bed of the Humber by the action of the tide and a certain amount of sediment brought down by the tributary streams which join the Humber some distance from its mouth.

The field or area to be warped must lie below the level of the water in the river at high tide. It is first surrounded by an embankment, after which the water from the river is allowed to flow through a properly constructed sluice in its bank, along a drain or ditch to the land which is prepared for warping. By a system of carefully laid channels the water flows gently over the land, and deposits its warp with an even level surface.

At the ebb of the tide the more or less clear water flows back again from the land into the main river with sufficient force to clean out any deposit which may have accumulated in the drain leading to the warped area, thus allowing free access of more warpladen water at the next tide. In this manner poor peats and sands may be covered with a large layer of rock soil capable of growing excellent crops.

The amount of deposit laid over the land reaches a thickness of two or three feet in one season of warping, which is usually practised between March and October, advantage being taken of the spring tides during these months.

The new warp is allowed to lie fallow during the winter after being laid out in four-yard " lands " and becomes dry enough to be sown with oats and grass and clover seeds in the following spring. The clover-grass ley is then grazed for a year or two with sheep, after which wheat and potatoes are the chief crops grown on the land.

Green manures are crops which are grown especially for the purpose of ploughing into the land in a green or actively growing state. The crop during its growth obtains a considerable amount of Green. carbon from the carbon dioxide of the air, and builds it up Manuring. i nto compounds which when ploughed into the land become humus.

The carbon compounds of the latter are of no direct nutritive value to the succeeding crop, but the decaying vegetable tissues very greatly assist in retaining moisture in light sandy soils, and in clay soils also have a beneficial effect in rendering them more open and allowing of better drainage of superfluous water and good circulation of fresh air within them. The ploughing-in of green crops is in many respects like the addition of farm-yard manure.

Their growth makes no new addition of mineral food-constituents to the land, but they bring useful substances from the subsoil nearer to the surface, and after the decay of the buried vegetation these become available to succeeding crops of wheat or other plants. Moreover, where deep-rooting plants are grown the subsoil is aerated and rendered more open and suitable for the development of future crops.

The plants most frequently used are white mustard, rape, buckwheat, spurry, rye, and several kinds of leguminous plants, especially vetches, lupins and serradella. By far the most satisfactory crops as green manures are those of the leguminous class, since they add to the land considerable amounts of the valuable fertilizing constituent, nitrogen, which is obtained from the atmosphere.

By nitrification this substance rapidly becomes available to succeeding crops. On the light, poor sands of Saxony Herr Schultz, of Lupitz, made use of serradella, yellow lupins and vetches as green manures for enriching the land in humus and nitrogen, and found the addition of potash salts and phosphates very profitable for the subsequent growth of potatoes and wheat. He estimated that by using leguminous crops in this manner for the purpose of obtaining cheap nitrogen he reduced the cost of production of wheat more than 50 per cent.

The growing crops should be ploughed in before flowering occurs; they should not be buried deeply, since decay and nitrification take place most rapidly and satisfactorily when there is free access of air to the decaying material. When the crop is luxuriant it is necessary to put a roller over it first, to facilitate proper burial by the plough.

The best time for the operation appears to be late summer and autumn. (J. PE.) Soil and Disease. - The influence of different kinds of soil as a factor in the production of disease requires to be considered, in regard not only to the nature and number of the microorganisms they contain, but also to the amount of moisture and air in them and their capacity for heat.

The moisture in soil is derived from two sources—the rain and the ground- water. Above the level of the ground-water the soil is kept moist by capillary attraction and by evaporation of the water below, by rainfall, and by movements of the ground-water; on the other hand, the upper layers are constantly losing moisture by evaporation from the surface and through vegetation. When the ground-water rises it forces air out of the soil; when it falls again it leaves the soil moist and full of air.

The nature of the soil will largely influence the amount of moisture which it will take up or retain. In regard to water, all soils have two actions - namely, permeability and absorbability.

Permeability is practically identical with the speed at which percolation takes place; through clay it is slow, but increases in rapidity through marls, loams, limestones, chalks, coarse gravels and fine sands, reaching a maximum in soil saturated with moisture.

The amount of moisture retained depends mainly upon the absorbability of the soil, and as it depends largely on capillary action it varies with the coarseness or fineness of the pores of the soil, being greater for soils which consist of fine particles. The results of many analyses show that the capacity of soils for moisture increases with the amount of organic substances present; decomposition appears to be most active when the moisture is about 4 per cent, but can continue when it is as low as 2 per cent, while it appears to be retarded by any excess over 4 per cent.

Above the level of the ground-water all soils contain air, varying in amount with the degree of looseness of the soil. Some sands contain as much as 50 per cent of air of nearly the same composition as atmospheric air. The oxygen, however, decreases with the depth, while the carbon dioxide increases.

SOIL CHARACTERISTICS

Scientists can learn a lot about a soil's composition and origin by examining various features of the soil. Colour, texture, aggregation, porosity, ion content, and pH are all important soil characteristics.

Colour

Soils come in a wide range of colours—shades of brown, red, orange, yellow, gray, and even blue or green. Colour alone does not affect a soil, but it is often a reliable indicator of other soil properties. In the surface soil horizons, a dark colour usually indicates the presence of organic matter.

Soils with significant organic material content appear dark brown or black. The most common soil hues are in the red-to-yellow range, getting their colour from iron oxide minerals coating soil particles. Red iron oxides dominate highly weathered soils. Soils frequently saturated by water appear gray, blue, or green because the minerals that give them the red and yellow colours have been leached away.

Texture

A soil's texture depends on its content of the three main mineral components of the soil: sand, silt, and clay. Texture is the relative percentage of each particle size in a soil. Texture differences can affect many other physical and chemical properties and are therefore important in measures such as soil productivity.

Soils with predominantly large particles tend to drain quickly and have lower fertility. Very fine-textured soils may be poorly drained, tend to become waterlogged, and are therefore not well-suited for agriculture. Soils with a medium texture and a relatively even proportion of all particle sizes are most versatile. A combination of 10 to 20 percent clay, along with sand and silt in roughly equal amounts, and a good quantity of organic materials, is considered an ideal mixture for productive soil.

Aggregation

Individual soil particles tend to be bound together into larger units referred to as aggregates or soil peds. Aggregation occurs as a result of complex chemical forces acting on small soil components or when organisms and organic matter in soil act as glue binding particles together.

Soil aggregates form soil structure, defined by the shape, size, and strength of the aggregates. There are three main soil shapes: platelike, in which the aggregates are flat and mostly horizontal; prismlike, meaning greater in vertical than in horizontal dimension; and blocklike, roughly equal in horizontal and vertical dimensions and either angular or rounded. Soil peds range in size from very fine—less than 1 mm (0.04 in)—to very coarse—greater than 10 mm (0.4 in). The measure of strength or grade refers to the stability of the structural unit and is ranked as weak, moderate, or strong. Very young or sandy soils may have no discernible structure.

Porosity

The part of the soil that is not solid is made up of pores of various sizes and shapes—sometimes small and separate, sometimes consisting of continuous tubes. Soil scientists refer to the size, number, and arrangement of these pores as the soil's porosity. Porosity greatly affects water movement and gas exchange. Well-aggregated soils have numerous pores, which are important for organisms that live in the soil and require water and oxygen to survive. The transport of nutrients and contaminants will also be affected by soil structure and porosity.

Ion Content

Soils also have key chemical characteristics. The surfaces of certain soil particles, particularly the clays, hold groupings of atoms known as ions. These ions carry a negative charge. Like magnets, these negative ions (called anions) attract positive ions (called cations).

Cations, including those from calcium, magnesium, and potassium, then become attached to the soil particles, in a process known as cation exchange. The chemical reactions in cation exchange make it possible for calcium and the other elements to be changed into water-soluble forms that plants can use for food. Therefore, a soil's cation exchange capacity is an important measure of its fertility.

pH

Another important chemical measure is soil pH, which refers to the soil's acidity or alkalinity. This property hinges on the concentration of hydrogen ions in solution. A greater concentration of hydrogen results in a lower pH, meaning greater acidity. Scientists consider pure water, with a pH of 7, neutral. The pH of a soil will often determine whether certain plants can be grown successfully. Blueberry plants, require acidic soils with a pH of roughly 4 to 4.5. Alfalfa and many grasses, on the other hand, require a neutral or slightly alkaline soil. In agriculture, farmers add limestone to acid soils to neutralize them.

SOIL USE

For most of human history, soil has not been treated as the valuable and essentially nonrenewable resource that it is. Erosion has devastated soils worldwide as a result of overuse and misuse. In recent years, however, farmers and agricultural experts have become increasingly concerned with soil management.

Erosion

Erosion is the wearing away of material on the surface of the land by wind, water, or gravity. In nature, erosion occurs very slowly, as natural weathering and geologic processes remove rock, parent material, or soil from the land surface. Human activity, on the other hand, greatly increases the rate of erosion. In the United States, the farming of crops accounts for the loss of over 3 billion metric tons of soil each year.

In a cultivated field from which crops have been harvested, the soil is often left bare, without protection from the elements, particularly water. Raindrops smash into the soil, dislodging soil particles. Water then carries these particles away. This movement may take the form of broad overland flows known as sheet erosion.

More often, the eroding soil is concentrated into small channels, or rills, producing so-called rill erosion. Gravity intensifies water erosion. Landslides, in which large masses of water-loosened soil slide down an incline, are a particularly extreme example. Wind erosion occurs where soils are dry, bare, and exposed to winds. Very small soil particles can be suspended in the air and carried away with the wind. Larger particles bounce along the ground in a process called saltation.

Soil Management

To prevent exposure of bare soil, farmers can use techniques such as leaving crop residue in the soil after harvesting or planting temporary growths, such as grasses, to protect the soil from rain between crop-growing seasons. Farmers can also control water runoff by planting crops along the slope of a hill (on the contour) instead of in rows that go up and down.

Soil faces many threats throughout the world. Deforestation, overgrazing by livestock, and agricultural practices that fail to conserve soil are three main causes of accelerated soil loss. Other acts of human carelessness also damage soil. These include pollution from agricultural pesticides, chemical spills, liquid and solid wastes, and acidification from the fall of acid rain.

Loss of green spaces, such as grassland and forested areas, in favour of impermeable surfaces, such as pavement, buildings, and developed land, reduces the amount of soil and increases pressure on what soil remains. Soil is also compacted by heavy machinery and off-road vehicles. Compaction rearranges soil particles, increasing the density of the soil and reducing porosity. Crusts form on compacted soils, preventing water movement into the soil and increasing runoff and erosion.

With the world's population now numbering upwards of 6 billion people—a figure that may rise to 10 billion or more within three decades— humans will depend more than ever on soil for the growth of food crops.

Yet the rapidly increasing population, the intensity of agriculture, and the replacement of soil with concrete and buildings all reduce the capacity of the soil to fulfill this need. As a result of an increased awareness of soil's importance, many changes are being made to protect soil. Recent interest in soil conservation holds the promise that humanity will take better care of this precious resource.

POOR SEEDLING EMERGENCE

Patchy crops reduce yield and profit. Poor seedling emergence is more of a problem under minimum tillage, because there is less soil disturbance and more stubble retention.

It is less of a problem under conventional tillage, because the seedbeds are finely worked and the seeds are planted into bare ground. However, rather than reject minimum tillage because of its problems, it is preferable to address those problems.

POSSIBLE CAUSES

Poor seedling emergence may be due to:

- Poor seed–soil contact
- Inaccurate seed placement
- A soil temperature that is too low or too high
- Soil insects or soil-borne disease
- Surface crusting after sowing
- Poor quality seed.

Poor seedling vigour and establishment after emergence may caused by the presence of compacted soil beneath the seedling. The soil may have already compacted before sowing, or the planting tine may have smeared the bottom and sides of the seed trench.

Planting Machinery

If poor seedling emergence is due to poor seed–soil contact or inaccurate seed placement, then adjusting your sowing machinery correctly will help to improve emergence.

Surface Crusting

Rain after sowing and before seedling emergence may form a surface crust that reduces emergence. Once seedlings have emerged, the problem of surface crusting is less. However, until the crop reaches full ground cover (and protects the surface from raindrop impact) surface crusting will continue to reduce infiltration.

All soils seal under raindrop impact—even self-mulching soils seal if they are bare. On some soils, the seal is weak or it cracks as it dries; it offers little resistance to seedling emergence. However, on other soils, the seal forms a crust on drying, and this crust prevents seedling emergence. Further light rain will help emergence by softening the surface; further heavy rain will hinder emergence by strengthening the surface crust. Clay soils can crust if the surface is sodic. The treatment for such soils is gypsum.

COMPACTION MANAGEMENT

A small degree of compaction can be beneficial. Soil that is too loose may not be good for plant growth: a loose seedbed dries out too quickly and gives poor contact between the seeds and the soil; seeds may not germinate. The optimum level of compaction depends on the soil types, the plant, and the irrigation management during the growing season. Too much compaction can cause yield reduction. To reduce soil compaction:

- Time mechanical operations carefully
- Reduce axle loads
- Keep livestock off wet cropping country
- Confine traffic to laneways
- Use low ground pressure (wide) tyres (with caution).

The method of compaction repair will depend on the soil's capabilities. A cracking clay that swells and shrinks behaves quite differently to a hardsetting, fine, sandy or silty soil. If a soil swells and shrinks on wetting and drying you can use that natural action to break up a compacted layer. If a hardpan has developed in a soil with little clay, you will need to consider other options.

Example:

- Biological repair—that is, using growing plants to break up a compacted layer or increasing the soil organic matter content
- Mechanical repair—that is, deep ripping or mouldboard ploughing.

BIOLOGICAL REPAIR

Biological repair means using growing plants to restore soil structure. It usually involves a yield penalty. The plants will suffer in restoring the soil structure, and will produce a lower yield. Repair of compaction should therefore work in with the season, the market prices, and your long-term plans.

Biological repair includes:

- biological ripping (using plants to dry and crack the soil)
- Biological drilling (using tap-rooted plants to 'drill' through acompacted layer)
- Increasing soil organic matter content.

Biological Ripping

The easiest way to restore soil structure in compacted cracking clays is to dry the soil, using any plants with a vigourous root system. Drying causes the soil to shrink and crack, allowing water, air and roots to penetrate through the cracks. The process is called biological ripping.

Biological Drilling of Non-Cracking Soils

In soils that don't crack, such as sandy, silty or loamy soils, 'biological ripping' is not an option. However, 'biological drilling' is possible. A strong tap-rooted plant such as lucerne can force its way through a compacted layer and leave root channels for subsequent plant roots to use. This is 'biological drilling'. The compacted layer must be moist: no root will enter dry soil. Depending upon how badly compacted the subsoil is, some roots will not be able to penetrate the compacted layer and the crop will suffer.

Increasing Soil Organic Matter

Some soils, such as fine sandy and silty soils, depend heavily on organic matter for the maintenance of their structures. Even cracking soils benefit from increased organic matter content. Increasing the organic matter of a soil content is a slow process. A pasture phase incorporated into your vegetable crop rotation is perhaps the most costeffective way to achieve this.

MECHANICAL REPAIR

Deep ripping and mouldboard ploughing are sometimes useful for quick results, but they are expensive operations. The initial success of the operation depends on the moisture content of the soil and the depth of tillage. Continuing success depends on managing traffic to avoid having machinery on the soil when it is too wet.

The Importance of Moisture Content

Soil moisture is important when you are working soil. A soil (particularly a clay) that is worked too wet will compact, while a silty or fine sandy soil that is worked too dry may powder. In both instances the soil will then structurally degrade. Before deep ripping, dig a hole to see if there is a compacted layer. If there is, check the layer's depth, thickness, and moisture content.

The severity of compaction will also influence your decision on deep ripping. If there is no compacted layer, there is no need to deep rip.

Deep ripping is expensive, and it could do more harm than good if the soil is too wet—a costly way of creating soil structural damage where none existed before! A compacted layer must be at the right moisture content to benefit from the ripping operation.

When you are cultivating clays (or when the compacted layer is clay) the layer must be dry enough to shatter, rather than smear. When the compacted layer is other than clay, some moisture is desirable so that the layer does not powder.

The depth and thickness of the compacted layer indicates how deep to set the ripper tine. To be effective, the tine should work just below the compacted layer. Dig a hole after a short run and check the effectiveness of the ripping operation.

❐

7.

Fungal Plant Pathogens and Symptomology

INTRODUCTION

Plant diseases are due mainly to the action of fungi, but some are caused by bacteria, a few by plasmodia, while others in which no particulate living organisms have, so far, been discovered, are believed to be due to virus infection. In plant viruses that have been studied in any detail the invisible agent multiplies with great readiness only in the living host from which the virus has been isolated in the form of a nucleoprotein. Other diseases, again, but of a non-parasitic nature, are attributed to physiological or to deficiency troubles, and probably all are influenced by the factors of the environment.

Including for the moment the bacteria with the fungi proper, one finds an enormous assemblage of species, at least as many as in the flowering plants, having few characters in common. They may be unicellular or composed of many cells, thread-like or forming sheets or masses, shapeless or beautiful in design, of limited or practically unlimited growth, of many colours. In the lowest group a cell wall- that standard criterion of the plant cell- may be present only in the sporing stage. Motility is not absent, either as the slow movement of plasmodia or the swimming of zoospores or the flagellar movements of many bacteria. In structure some approach closely to the green algae and, as the name Phycomycetes or algal-fungi reveals, it was long a popular suggestion and is still sometimes upheld that they were derived from algae; the best modern view, however, is that they arose from colourless unicellular protozoa, and it is even maintained that, since the green algae probably arose from similar protozoa that had acquired chlorophyll, and since the rest of the green plants have developed along that path, it is logical to regard the fungi as a distinct phylum, neither animals nor plants but co-equal in their origin with the green organisms forming the plant kingdom. Normally, however, they are treated as plants, included in botanical works and courses, and they may be accepted as such. The chief characteristic of the fungi is the absence of chlorophyll, even in those members which happen to be coloured green. This carries with it a profound difference in mode of life from other plants, for they take their food in the form of organic matter, whereas the chlorophyllous plants obtain their carbonaceous food from the carbon dioxide of the air, through the energy of sunlight. The flowering plants, ferns, mosses, and algae build up their food out of dissolved inorganic salts from the soil or water, together with the constituents of the air.

The fungi are unable to do this; they are incapable of photosynthesis and so cannot make carbohydrates from carbon dioxide and water; they must have organic matter or food already prepared by having been built up into the bodies of plants and animals, or having once formed part of these, like jam or leather. They are scavengers, without whose cleansing aid in decomposing organic matter, the earth would soon become unfit to sustain the higher forms of life.

BACTERIA

The bacteria really have very little in common with the fungi proper, except, in general, their mode of feeding. They are extremely minute, usually consisting of a single cell or a collection of more or less independent units. Sometimes these units are arranged in branching threads which are hard to distinguish from those of the filamentous fungi.

The fungal thread, however, is normally a part of the one individual, whereas a bacterial thread consists of a chain of individuals more or less loosely joined together. The branching is also different, the fungi having true branches, while in a bacterial column the spurious appearance of branching is due to imperfect separation of broken parts of the chain, which continues to give off new individuals above and below the break.

The most important distinctions between the two groups are in the methods of growth and reproduction. Fungi normally grow only at the ends of the threads which compose their bodies, and multiply by forming spores at the ends of, or within special threads.

Bacteria, on the other hand, have not got this purely end growth, and usually multiply by each cell simply' dividing into two cells, which frequently separate from each other and become independent.

There are other less important differences as in the nuclei, cell wall, and so cl Being smaller and lighter than most spores of filamentous fungi the bacteria are even more efficient than the former in reaching and decomposing organic waste; as f§ well known, they were the chief subject of Pasteur's convincing demonstration against the old-fanged notions of spontaneous generation.

PLASMODIOPHORALES

The Plasmodiophorales are an aberrant group having apparent affinities with the chlorophyll-free mycetozoa or myxogastres, organisms close to the dividing line between animals and plants. They are now usually placed near or in the primitive Chytridiaceae. They have a power of spontaneous creeping movement not limited to the zoospores. Instead of a germ-tube the spore gives rise as a rule to motile swarm cells which fuse to form a small mass of naked protoplasm endowed with movement resembling that of an amoeba.

This forms a multinucleate' plasmodium' or plasmatic body still without a wall and usually constituted by the amalgamation of several individuals. In at least some of the genera the swarm cells function as gametes and the plasmodium arising from the zygote forms sporangia, so that there are well- marked sexual and asexual phases in the life-cycle.

Two important crop parasites binning to this small family occur in the British Isles, namely Plasmodiophora brassica and Spongospora subterranea, causing' club root' of cruciferous crops, and' powdery scab' of potatoes, respectively.

THE FUNGI PROPER

All known fungi, with few exceptions, originate from spores, bodies comparable with, though not strictly similar to, seeds, for they have no preformed embryo or germ. These spores germimtte, requiring for the purpose much the same conditions as ordinary seeds.

The result of germination is usually the protrusion of one or more fine filaments known as germ-tubes. These grow and give off branches, the resulting branched thallus constituting the' mycelium' or vegetative part of the fungus, as distinct from the reproductive part. Each individual filament of the mycelium is termed a 'hypha'. In a limited number of primitive fungi the vegetative part consists of a single cell, usually round' or oval, and this may become transformed into a reproductive cell with little change in shape, the life-history being a very simple one.

THE VEGETATIVE STRUCTURE OF FUNGI

THE MYCELIUM

At an early period in its growth the germtube and the hypha which arises from it usually become segmented into a row of cells by the formation of transversely walls each, when young, being perforated by a central pore sufficiently wide to allow protoplasmic granules and even nuclei to pass from cell to cell. Each cell is a hollow structure, bounded by walls of a transparent pliable material, which is sometimes cellulose as in the higher plants, but more often is a complex nitrogenous substance which resembles chitin, the material which forms the covering of insects.

Within the walls, the cell is filled with a watery jelly, protoplasm, a material mainly protein in constitution, found in all living bodies and, in fact, the actual living matter itself. The cell cavity is not uniformly filled. Here and there little spaces occur and are known as 'vacuoles'; they are usually filled with water containing dissolved salts, sugar, and so forth, of the cell sap. Here and there also condensations of a peculiar kind of protoplasm form, which are termed nuclei. One or more of these nuclei may occur in each cell; two are habitually found in certain phases of the life-cycle of Ascomycetes and Basidiomycetes, playing a part in the complicated sexual reproduction of these fungi which will be further discussed below.

In the lower fungi it is common, or even in some families habitual, to have no transverse walls in the active hyphae, which are' coenocytic "that is to say, composed of a single large cell with many nuclei. The nuclei are the bearers of the hereditary characters of the cell and also play an important part in regulating its metabolism.

GENERAL PRINCIPLES

The plasm with its nuclei but many other substances, such as reserves of sugars, fats, glycogen, crystals, resin, pigments, and so on. They are of many shapes, but on the whole the spores preserve considerable uniformity within the species. The prodigality of nature in its reproductive schemes is illustrated by the fact that a single giant puffball has been estimated to bear some 100 million spores. It is little wonder that every substance suitable for colonisation by fungi is reached by air currents charged with spores, unless special precautions are taken to exclude them.

The body of the simpler fungi usually consists of branching hyphae only. It may be short or long lived. The' fairy ring' fungi, of which Marasmius oreades is the commonest in England, are the best examples of perennial mycelium, as they may survive for as many as four hundred years, producing a crop of toadstools each year. In many cases unions termed' anastomoses' occur between neighbouring hyphae which may be from the same or different spores of the fungus concerned. These form bridges across which nutrient material and even sometimes nuclei pass from one hypha to another. Most fungi, however, have that part of the body which is involved in spore-bearing of more complex structure, formed of hyphae of special shape, or of bundles of hyphae, or of an interwoven tissue derived from hyphae by their repeated division into short cells. This tissue is termed' plectenchyma 'or, since it resembles to some extent the parenchymatous tissues of higher plants,, pseudoparenchyma'.

CHLAMYDOSPORES AND GEMMAE

Hyphae of the ordinary vegetative mycelium may also become modified. Some of their cells may become provided with thickened walls, able to withstand adverse conditions, and may even assume a characteristic shape and colour. They are known as chlamydospores and, like true spores, are usually microscopic.

As they become separated from the mycelium when mature, and can germinate like spores, they playa part in the dissemination of the species which, on account of their durable nature, may be very important. Scarcely to be separated from the chlamydospores, except in being less durable, are the cells termed gemmae. They are usually thin-walled and contiue to bud out cells similar to themselves instead of giving new hyphae.

Under certain conditions the mycelium may be almost wholly transformed into these masses of rounded cells. Indeed, the familiar group of the yeast fungi may be regarded as an extreme case of the reduction of the mycelium to small gemmae endowed with a remarkable facility for budding out similar cells so as to form colonies which readily break up into their component units. This budding mycelium is frequently associated with fermentative activity.

SCLEROTIA

Much larger and more complex masses of vegetative hyphae-than those producing chlamydospores may become united into, sclerotia', sometimes measuring several inches across and as hard and heavy as a stone, but still formed wholly from vegetative hyphae variously thickened and divided so as to lose their filamentous character, and become pseudoparenchymatous. The nests of white ants in hot countries often contain numerous rounded to spindle shaped stalked sclerotia belonging to the Ascomycete Xylaria injuries while a sclerotium which may be as big as a toy football is formed by the basidiomycete Polyporus mylittae in Australia, where it is known as 'black fellows' bread', as it is eaten by the aborigines.

Several other idiomycetes also form large sclerotia. Similar but much smaller types are known in both hot and temperate climates, the commonest in the British Isles being the often irregularly shaped sclerotia of Botrytis or Sclerotinia, of black-dot of potato and tomato Colletotrichum atramantarium and of Corticium solani, the cause of, black scurf' of the potato, etc. The round sclerotia of Sclerotium rifles, resembling very small seeds, are derived from a parasite destructive in the United States and elsewhere, while the still smaller, almost microscopic, Sclerotium stage of the pycnidial fungus Rcrophomina phase oli is ubiquitous in India and known in several other hot countries.

In some sclerotia, such as those of Botrytis or Sclerotinia, a hard outer rind or cortex of thick-walled, often dark cells, with small cavities, may be distinguished from a softer pith. Besides being useful stores for reserve nutrient material, sclerotia are particularly well suited to withstand unfavourable extremes of temperature, dryness, and the like, and are valuable in carrying the organism over long periods of adverse conditions.

RHIZOMORPHS

Visible strands or cords of mycelium can often be seen on turning over heaps of leaves or at the base of puffballs and other large fungi. The simplest of these are generally white, and beyond being intertwined or sometimes laterally united they do not appreciably differ from an ordinary bundle of vegetative hyphae. More complex strands, known as, rhizomorphs "are found in the vegetative thallus of some fungi. In the less differentiated types of these, ordinary filaments unite with specially tough hyphae and sometimes with hyphae adapted for the conduction of nutrients to constitute a part of the plant fitted for extensive exploration in search of new colonisation grounds. In the dry rot fungus Merulius lacrymans they have been found passing through the mortar of walls to reach fresh structural woodwork on the other side.

The more highly differentiated rhizomorphs are represented in the common parasite of tree roots Armillaria mellea. These have been likened to purplish-brown or black leather shoestrings, by they are often much branched, rather uneven in diametre, smooth and shiny when fresh, and luminous when young.

They consist of an outer black, rather brittle rind, composed of brown, thick-walled pseudoparenchyma, and a light-coloured tough flexible pith in which the filamentous character of the hyphae is largely preserved. Though they grow normally by definite apical growth, keeping their root-like form, they can spread out into sheets of mycelium in certain positions, especially when they pass under the bark, or they may burrow deeply into the tissues of the root, forming tubular structures in the long axis of the root; these are still differentiated into a harder cortex and a loosely interwoven pith, and grow only at the apex.

Similarly embedded rhizomorphs are formed by the Ascomycete, a widely distributed: Ingus causing root rot of tea and other crops. In the more-serious of the root diseases of tropical plantation crops, such as rubber and tea, the rhizomorphs, which are the chief organs of spread, remain on or in the roots or other buried timber and even in Armillaria mellea their ability to grow away from such supports and food-base is limited, and extension through firm soil, other than along wood, is almost negligible. Their function in the parasitism of the last-mentioned fungus is discussed in a later chapter.

Allied to rhizomorphs are the raised black lines or narrow bands running longitudinally on the surface of roots attacked by Armillaria mellea, each representing the free margin of a dingy white sheet of mycelium which disrupts the wood, often but not invariably in a radial direction. The xylostromata are capable of growth at their free edges and of causing infection when a diseased root comes into contact with a healthy one. They are much less often seen in temperate than in tropical forms of the fungus, in contradistinction to the rhizomorphs, which are common in temperate, but may be hard to find in tropical countries.

The xylostromata arise by proliferation from a narrow band of deep brown cells which form sharply defined black lines in the tissues invaded by Armillaria mellea, These bands are sections of continuous sheets crossing all the tissues of the host from near the surface to the pith. They begin by a development of thin-walled, swollen, colourless, bladder-like cells completely filling the host cell cavity. The bladder cells soon thicken their walls and turn brown.

When fully developed, the black line is composed of a dense mass of angular, often rather small cells of a deep brown colour, and this colour also stains the walls of the host cells. The result is a continuous layer surrounding an area rotted by the fungus, and it has been suggested that it encloses a mass of fungal tissue, mixed with elements of the host, which is protected from adverse influences and to which the term 'pseudosclerotium' has been applied. Black lines of this type occur in the tissues invaded by various parasitic and saprophytic species of Ustulina, Xylaria, Hypoxylon, and the like.

STROMATA

In many fungi, after a period of vegetative growth in the filamentous condition, the mycelium condenses here and there into pseudoparenchymatous masses termed' stromata "which are formed generally on or just below the surface of the substratum in which the fungus is growing.

These differ from lerotia only in having usually a less regular shape and a margin in more frequent communication with the filaments of the rest of the thallus. They often, indeed, as in Rhytisma, only form flat crusts covering the deeper hyphae from which they have developed. They commonly constitute a transitional stage between the purely vegetative and the reproductive parts of the fungus, and the term, storm is sometimes restricted to this pre-sporing stage. It is difficult, however, to accept this restriction, as stromatic crusts are common that seem to have no association with spore bearing.

The parts of the fungus plant so far considered are chiefly concerned with nutrition. The more specialised rhizomorphs, sclerotia, resting and budding cells, take part also in propagating the species, but this is 'vegetative' propagation, on a par with propagation by cuttings, buds, suckers, and the like, amongst higher plants. The reproductive system proper is usually quite distinct and well defined.

REPRODUCTION OF FUNGI

Sporophores

As the vegetative system is usually buried in the substance on which the fungus feeds- in the soil, in rotting organic matter, or in living plants- and as most spores should be formed in free access to the air to secure their dissemination, it is evident that some special arrangement must be made to bring the spore-forming parts where they will be exposed to the elements.

This is done usually by the development of structures known as sporophores raised vertically above the level of the general mycelium. The sporophore may consist either of a single hypha or of bundles or complex masses of hyphae and it bears on some part, usually towards the top, the swore of tnfungus. The hyphae of which it is corpuscle ordinarily differ from those of the vegetative mycelium in one or more characters, such as their vertical position, limited growth, or special structure and as "they may be arrange a in systems of great complexity, an almost endless series of different types of sporophore exists.

The simplest Iotina-commonly 'in riiany of the lower fungi and in those moulds that are familiar on damp or rotting organic substances and that belong to several different families of the fungi. In some of these, particularly such as have a considerable development of aerial mycelium in addition to the hyphae buried in the substratum and some of those that grow in water or in soil, special sporophores are almost or quite dispensed with, as being unnecessary to secure free dissemination.

The spores are formed at the tip of, or laterally on, ordinary hyphae, or, in the aquatic species, in special cells on the hyphae, and are liberated in the air, the water, or the soil as the case may be. More often, however, specialised sporophores are formed, and in the simplest cases these are single hyphae called sporangiophores, or conidiophores, often of distinct shape and size, which grow out into the air, and bear the spores on the tips or sides or on lateral branches.

In other cases, clumps of hyphae growing close together rise into the air. Sometimes these merely remain near together without lateral union, sometimes they unite more or less closely into solid columns.

Sporophores aggregated into crusts or layers are produced in many fungi. Sometimes these are formed on the surface of the substratum in which the vegetative mycelium grows, sometimes, especially in parasitic and other fungi that grow in the stems and leaves of plants, just below the surface, which is eventually ruptured so as to expose the spore-bed. In either case, the layer of sporophores may arise directly from the mycelium below, or there may first be formed a stromatic tissue composed wholly or in part of pseudoparenchyma. Common examples of spore-beds formed below the surface of the plant are the, sari' of the cereal and other rusts and the 'acervuli' of many of the Ascomycetes and of the 'Fungi Imperfecti'.

When formed on the surface, there may be a very large mass of stromatic tissue intervening between the vegetative mycelium and the layer of sporophores. Such compound sporophores in the form of superficial stromatic crusts or tubercles may be found commonly on pieces of old wood, dead branches of trees, and the like, in damp situations. Sometimes the edge of the crust becomes free and curves over to form a sort of bracket. This is the first stage in a process which has as its purpose the lifting of the spore-bearing part of the sporophore above the object on which it is growing. In more advanced types, so much of the crust may leave the surface that only a narrow band or stalk remains to join it to the latter. In this way the bracket fungi found on trees and old wood, and the laterally attached toadstools, are formed. Finally, the whole spore-bearing part of the fructification may be raised from the surface on a central stalk and may be expanded into a cap at the top, as in the well- known mushrooms, or may remain as a simple or branched, vertically elongated body of various shapes.

In another type of compound sporophore, the spore-bearing hyphae are enclosed in roundish or cup-shaped or flask-shaped receptacles. To a common form of this type the name pycnidium 'is applied when it bears spores on the or immersed in it) e.g. Epichloe, Claviceps and opening to the surface by the narrow neck through which the spores are expelled when-ripe.. Sometimes they are formed close together but free from each other as in Nectar and Ophiobolus, sometimes in groups immersed in a 'common stroma as in Claviceps and Epichloe, and, as in these two types, the stromabas in some cases become eleyated or vertically elongated so as to raise the spore receptacles above the level of the sub stratum.

In most compound sporophores, the spores are not produced on all parts of the organs. In the mushrooms, for instance, they are formed only on the surface of the gills found on the under side of the expanded 'cap'. To the fertile part of the sporophore, i.e. that part which actually bears spores, the name 'hymenium' is given.

In many cases, where the spore-being hyphae are so numerous as to form definite layers or a hyrpenium, certain sterile hyphae usually of characteristic form, occur intermingled with the fertile ones. These are frequently known as, paraphyses' and are especially common in the Ascomycetes and Basidiomycetes. The basidia of the H ymenomycetes are often interspersed with sterile cells or 'cystidia'.

From the above account of the fungus body, it will be seen that an ordinary fungus consists of a vegetative part, usually formed of thread-like hyphae buried in the substratum, find of a reproductive part which is sometimes not distinct from the vegetative except in bearing spores, but is more usually modified to form a simple or compound sporophore. As the production of spores provides for a new generation of the fungus, one may consider that the life-history of any species begins with the germination of the spore and ends with the development of new ripe spores, or in the case of those fungi which have more than one kind of spore with the development of the final or, as-it is often termed, 'perfect spore form, which is usually the result of; sexual process at its equivalent.

FORMATION OF SPORES

The spores of fungi present a wonderful range of variation, there being hundreds of types differing in size, snaps, structure, and more of formation. The one fungus may have several different kinds of spore during the course of its complete life history. The vast majority differ sharply- from the vegetative cells of the mycelium, but in any single species each kind of spore remains remarkably constant. Two main types may be distinguished: a 'perfect' form which is the result, immediate or delayed, of a sexual process and an 'imperfect' or 'asexual' from to which the term sporangium or conidium' is usually applied. When the thallus which bears self capable of sexual fusion is distinct from that carrying-asexual spores, the former is termed the gametophyte and the latter, the spordphyte. The best example of this alternation of generations amongst the fungi' is found in the genus Allomyces but it is much less clear cut in fungi than in such plants as mosses and ferns.

SEXUAL REPRODUCTION

The perfect form of fructification differs widely in the three main groups of fungi, Phycomyceies, Ascomycetes and Basidiomycetes. In the Phycomycetes it is a thick-waned, well-defined cell, to which the general term 'zygote' is applied in some of the more primitive forms, while in the Oomycetes the zygote is termed apoospore and in the Zygomycetes, a 'zygospore'. A zygote is simply the end product of the fusion of two cells with sexually differentiated nuclei which also fuse. These cells are known as 'gametes' and in some of the Chytridiaceae are motile; they may be outwardly alike, so that a IV-ale is not to be distinguished from a 'female' gamete.

An oospore is a cell resulting from the fertilisation of a larger female' cell or 'oogonium' by a smaller male cell, the' antheridium "both borne on hyphae of the mycelium. A zygospore results from the union of the tips of two branches of the mycelium, often quite similar to one another. In all these cases fertilization is completed, as usual, by the fusion of the nuclei from the two uniting gametes to form the zygote. At some stage prior to fusion the nuclei undergo a 'reduction 'division, which results in the halving of the number of 'chromosomes' and gives rise to what is termed the I, haploid' phase of the life-cycle.

When two haploid nuclei come together in a cell, the dikaryophytic, dikaryo-, or dikaryotic phase is initiated; fusion of the two haploid nuclei restores the original number of chromosomes, half of which are now of different origin from the other # half, and the I resulting nucleus is termed 'diploid'.

The process directed to bringing together I the two haploid nuclei in one cell is sometimes called 'diploidisation'.

In the Ascomycetes the perfect stage is the 'ascus', a saclike, roundish or elongated cell containing' ascospores "which is sometimes immersed without orderly arrangement in a stroma or in meshes of the mycelium but usually lies side by side with other asci and, perhaps, paraphyses to form a definite humanities. In some forms the ascus is the result of the fertilizations of a female cell, the' ascogonium.', through a process or 'trichogyne' which contacts an antheridium-like male cell.

Instead, however, of the two nuclei uniting as in the oogonium,,they- may divide several times and a male and a female pair pass into each of a group of cells which grow out from the ascogonium; from one or more of these cells the asci 'arise, each ascus still having a male and a female nucleus. Fusion of.the two nuclei occurs in the young ascus and the ascospores arise by the division of this sexual nucleus. Owing to their relative sizes the nuclei often have to lie in single file in the ascogenous hyphae: and an arrangement known as the 'crosier' of nuclei, male and female, in the penultimate cell, from which the ascus arises. The clamp Connection of the Hymenomycetes, to be described below, serves the same purpose.

There is still some disagreement as to the number, location and significance of the nuclear fusions preceding the formation of ascospores and the details of the sexual process are as yet uncertain in some forms. In some cases a fusion of the male and female nuclei has been observed in the ascogonium, in addition to the fusion in the young ascus, and two reduction divisions following this double fusion have been described.

In the unicellular yeasts the single cell may become converted as a whole into an ascus, the formation of which may be preceded by union with another cell by means of a short bridge across which the nucleus,of one cell passes to face with that of the other. In cultures of-these fungi some cells are haploid, some diploid. The great bulk of the Ascomycetes are contained in' two main groups, the Pyrenomycetes in which the sporophore is a flask-shaped perithecium often immersed in the tissues or in a stroma and opening by a narrow neck and the Discomycetes in which the hymenium lines the inner surface of a cup-shaped sporophore or apothecium. In the Pyrenomycetes, which are the most numerous and diverse group of Ascomycetes, the hymenium lines the inner surface of the perithecium and the spores may se extruded while they are still enclosed in their asci or after they have been set free into the' cavity of the sporophore.

In the heterogeneous group, the Plectomycetes, the powdery mildews, are furnished with a characteristic type of fructification; the cleistocarp or cleistothecium, which is not provided with opening and the ascosporesare only liberated after the protective wall ruptures or decays. The Exoascaceae are included in this group although the asci are not enclosed in a special body at all, but like Taphrina deformans the cause of peach-leaf curl are exposed on the surface of the host once they break through the cuticle at maturity. The typical number of spares in an ascus is eight but there may be fewer or more, when the number is increased by division of the original nuclei, usually to a higher power of 2, as 16, 32,-.64, 128, but sometimes irregularly from suppression, of one or more spores.

In many of the higher Ascomycetes the ascogonium which eventually gives rise to the ascogenous hyphae appears often to be fertilized by microconidia. The oidia may be borne laterally on small awl-shaped or verticillate sporophores and in certain fungi as in the genus Sclerotinia do not function as ordinary spores as they do not give rise to a mycelium. When set free in Sclerotinia they adhere to cells either in special receptive bodies raised above the level of the stroma or contained in the sclerotia which are characteristic of these fungi. Fusion presumably results in the transference of a nucleus from a hypha in one mycelium to another, and apothecial development follows.

Spermatia borne in special spermagonia have long been known to occur in the life-cycle of many rusts. Pyrenomycetes and Discomycetes and to have no apparent function as sprees capable of reproducing the species. In a few cases they have also been found fusing with cells of the primordia of pedthecia, and though the evidence is largely circumstantial, it is probable that both oidia and spermatia ordinarily function as haploid' male' cells.

Turning to the Basidiomycetes, a still greater complexity is found in the processes which seem to.serve the ends of a sexual union. In place of the ascus of the Ascomycetes, the 'basidium' marks the perfect stage in the life-cycle. The term' promycelium' is also applied to the basidium in this group. It is a round, club shaped or baton-shaped body in which the gametic nuclei fuse. Subsequently it gives out a group of spores on short outgrowths from its surface. In the wide sense, the Basidiomycetes include the smuts and rusts as well as the mushrooms and bracket fungi, and the different groups vary considerably in the details of their perfect stages since the' sporidia of the two former groups do not arise immediately from the zygote.

The sporidia of the smuts are borne typically on a promycelium which resembles a stout germ-tube arising from the smut spore. In the majority of the species the promycelium is divided by transverse septa into four cells each of which contains a single nucleus. Usually a single sporidium or hyphal branch is produced from each promycelial cell, but sometimes two or more are developed; the promycelial nuclei do not pass out into the sporidia but divide, and each of the daughter nuclei enters a sporidium.

In some cases this nuclear division may be repeated several times so that crops of sporidia arise directly from a promycelium. Sometimes the four cells of the promycelium use in pairs by means of small buckle-like outgrowths through which a haploid nucleus passes from one cell to the other. Thus the binucleate, dikaryotic condition in the smuts may arise from the union of two sporidia or hyphal branches or from the union of two promycelial cells. In the genus Tillejia, the nuclei usually pass into the sporidia or hyphal branches, none remaining in the promycelium which in this type usually remains uniseptate.

In Tilletia the primary sporidia formed at the tip of the promycelium become binulceate by fusion in pairs; they give rise to fine hyphae on which are borne, secondary sporidia which are usually also binucleate but on germination give uninucleate and presumably haploid sporidia. These secondary and tertiary sporidia are considered by some to be the true basidiospores. The smuts are parasites, and in a good many species it is the binucleate germ-tubes which enter the host plant, the cells of the mycelium in the host usually again appearing binulcleate just before the smut spores form, when the two nuclei fuse, transforming the smut spores into diploid zygotes.

In some species the fusion hyphae are formed within the tissues of the cost and there is some uncertainty as to the details of the change from the haploid, to 'the diploid phase in these. The completion of the sexual process from the time the two nuclei are brought together in a cell until their fusion in the young spore may take several months. The sporidial stage can be readily grown in culture and experiments with cultures derived from single sporidia have established that fusions will only occur between certain pairs and that the four promyceliaf cells, usually represent two males and two females so that a simple form of heterothallism exists., Since in some species more than one pair have been found capable of fusing it is necessary to recognise the existence of more than two kinds of sexually functionable gametes in fungi or else to use another terminology. Thus some speak of the 'A', 'B', 'C', etc., 'genders' instead of sexes, or of compatible and incompatible groups. At least nine such groups appear to be recognisable in maize smut.

It has Jong been known that the rusts have a dikaryophytic phase in which two nuclei occur in each cell. This commences as a rule somewhere in the primordia of the acedia and terminates in the ripe teleutospore where the two nuclei fuse to form the zygote. Rusts are amongst the most pleomorphic of the fungi, the full life-cycle often possessing no less than five spore stages. The basidium arises from the germination of the teleutospore—a resistant spore- stage. Homologous with the smut spore and haploid phase is initiated during the process of nuclear division in the promycelium, and which normally results in the production of four sporidia one from each of the four cells into which the promycelium becomes divided by trans verse septa.

As the four sporidia belong to two sexes. The haploid sporidia infect the host and cause the development of immersed flask-shaped spermagonia containing masses of tiny spermatia which are extruded in a drop of sweet 'honey dew' or nectar through the neck or pore to the surface of the plant.

Here they are visited by flies or other insects which carry them to the openings of other spermagonia where they adhere to certain emergent hyphae in or at the margin of the spermagonium.

The next stage is difficult to follow and seems to differ in different species. In some, at least, the nucleus of the spermatlum appears to be able to pass back through the hypha with which it fused to reach the deeper levels of the haploid mycelium developed from the sporidium. The general outcome is the establishment of a binucleate condition of certain haploid cells at the base of the future sporophore which develops in the vicinity of the spermagonia, often on the opposite side of the leaf. This is the aecidium, bearing binucleate aecidiospores in chains.

From the dispersed aecitliospores arises a parasitic mycelium, the paired nuclei in the cells of which are maintained by conjugate division. It is this dikaryophytic mycelium that gives rise to binucleate uredospores, often in repeated crops of uredosori in such quantity that the yellowish or rusty spores colour the surface of the soil under the infected plants. Ultimately teleutosori replace the uredosori and the paired nuclei at last fuse in the mature teleutospores.

The rusts frequently overwinter in the teleutostage, producing their promycelia when the teleuto spores germinatein the spring. In a considerable number of species the haploid sporidia from the promycylium are alliable to infect the species of host plant on which the uredo- teleutospore stage develops, but attack some other species, often far removed systematically from the first and cause the development in it of a haploid vegetative mycelium leading to the aecidial stage while the binucleate aeeidiospores carry the fungus back to the original host. This which is known as heteroecism will be further discussed in a later section and seems to be a step in advance along the path already entered by the smuts, where the haploid germ-tube may be devoid of parasitic capacity.

In the Basidiomycetes proper the conception of sexuality must be further stretched if it is to cover all the known cases. Not only are the four sporiqia or basidiospores that are normally borne on four sterigmatabn the free end of the basidium often only able to form thalli that can fuse in pairs, which are mutually interested, but isolation of sporidia from other sporophores or other regions may go on adding almost indefinitely to the number of sexually functional gametes found. In these fungi the haploid spore stage may be represented by uninucleate oidia horne laterally on the mycelium, which may fuse together or with the end cells of hyphae composed of uninucleate cells. The result is a binucleate cell and the binucleate condition in subsequently developed cells of the terminally growing hypha is maintained by a curious dikaryophytic phase of the mycelium, develops around the September. I turn formed when the end cell of the hypha divides. The clamp begins as a protuberance in the side of the end cell just above I where the new septum will form Both nuclei move to this point and one enters the protuberance. Both then qivide while the septum separating the parent cell into two develops between the pair arising from the nucleus which did not migrate into the. protuberance.

The clamp completes its development by uniting at its tip with the cell below the new septum, and a nucleus from the pair formed in it passes to each of the two cells of the hypha. Thus the binucleate cells always contain a nucleus from each of their two components. Eventually the two fuse in the basidium and the perfect stage in achieved. Some times the dikaryophytic mycelium arises from the fusion of haploid mycelia derived from oidia of opposite sex as well as from haploid mycelia of.opposite sex derived from the germination of the basidiospores.

These haploid mycelia are without clamp connections and are sometimes termed primary mycelia as distinct from the' secondary, mycelia of the dikaryophytic phase. Oidia are some times developed on secondary mycelia, when they are usually binucleate and function as ordinary conidia. There appear to be many interesting variations in the diploidisation process, even to the transfer of nuclei from a diploid to a haploid mycelium, but it would be outside the scope of this survey to follow these or the debates to which they have given rise as to the physiological significance of the nuclear behaviour.

There have been several reports of clamp connections in the smuts where the paired nuclei cell is less rigidly preserved during parasitism in the rusts. It seems that both in smuts and Hymenomycetes the partnership of the two nuclei is liable to dissolution. In Urocystis cepulae the parasite of onion smut, the cells remain long uninucleate in onion tissues, becoming dikaryophytic just before sporulation.

HETEROTHALLISM

In all the main groups,of fungi it Lias been found that individual thalli or vegetative bodies are not necessarily identical in all their attributes in any given species. The difference to which the term' heterothallism 'is applied relates to sexual behaviour; in 'homothallic' forms thalli derived from a single spore can give the perfect stage, whereas in heterothallic species.

The perfect stage is only produced when union occurs between the cells or gametes of certain pairs of thalli. Thalli that can contribute to the development of the perfect stage are sometimes termed 'plus' and 'minus' thalli but it is scarcely possible to restrict the conception to a single pair of signs. It is evident that heterothallism is a means of securing outbreeding, a matter the biological value of which is indicated by the many well-know adaptations directed to secure the cross-fertilization. The apparent capacity of the different strains inheterothallic fungi to maintain themselves indefinitely in the imperfect stage, so that there is no way of knowing, that they have the ability to produce the perfect form unless chance or deliberate experiment reveals it, suggests that some at least of the great mass of the Deuteromycetes or 'imperfect fungi' may simply he strains that have lost their mates.

An alternative suggestios that nutrition rather than sexuality forms the basis of heterothallism has found some support. In this nuttitive theory each of the various strains is thought to possess a gene or factor absent from, the others; enabling it to utilise some essential food material. Union of compatible strains brings ability to use all that is required.

ASEXUAL REPRODUCTION

Out of the multiplicity of asexual spore forms found in the fungi certain main types may be mentioned. The 'sporangia' of the Phycomycetes maw contain unicellular 'zoospores', the animal-like attribute implied in the name being their power of swimming about in water by the movement of one of two 'cilia' or 'flagella', oarlike protrusions of the naked protoplast, attached terminally or laterally. After this free-swimming period the zoospore comes to test, rounds off, usually becomes provided a cell wall, and then germinates by a germ-tube or by a short process which penetrates the host cell. Germination sometimes results in the liberation of a second zoospore on which the position of the flagella may now be different, apical or lateral, from their previous position, so that the period of free-swimming life is pro longed.

A transitional stage in the adaptation of the aquatic fungi to life on land is, perhaps, to be observed in the Peronosporaceae and some of their allies. In many of these the sporangium often fails to produce zoospores but germinates directly by one or more germ-tubes, just as if it were wholly transformed into a single spore. In the Zygomy cetes, unlike the aquatic families of the Phycomycetes, the spores borne in the sporangia are not motile but are suitable for passive dissemination in the air or water; these are amongst the commoner land moulds and include the well known Mucoraceae such as the bread moulds and others found on decaying organic matter in the soil and elsewhere. Most mucors have mucilaginous spore heads and are not readily dispersed until the head breaks up in water or in the soil solution and the individual spores adhere to soil particles and the like and are dispersed on.these.

In some of the other genera these dried minute unicellular spores, when liberated by the disintregation of the sporangial wall, become air borne and are blown about as dust. Pilobolus has the whole sporangial head shot by a water jet generated in the sporangiophore to a distance said to be sometimes as much as a yard.

Contrivance called a 'clamp connection. This, whose present makes the term 'conidium' is applied, and this name is also often given to those sporangia of the Phycomycetes which germinate directly by germ-tubes; they are, nevertheless, potential sporangia in this group. The conidia of the Ascomycetes may be composed of one or many cells, uni- or plurinuceate, and of the most diverse shapes from simple rounded or oval forms, to branched bodies, sometimes furnished with slender cilium-like antennae from the end cells.

In the simplest case a conidium is merely the end of the fertile hypha or of a small stalk from the end or side of the latter. The spore is cut off by a septum from the parent hypha and may be called a thallospore. Very often, instead of a single spore being thus formed, a chain of spores arises from the sporophore or conidiophore. These chains may be developed in either of two ways; most commonly, each new spore of the chain is successively cut off from the tip of the parent hypha so that the end spore of the chain is the oldest; less often, the first-formed spore buds out another from its tip, and this in its turn another... so that the end spore is always the youngest. As there may be more than "one bud from the tip of a spore, the chance of the second type are often branched.

After a conidium or a chain of conidia is formed on "the end of the conidiophore, the latter may continue to grow from a point just below the tip. The new growth usually tends to preserve the same direction as the original stalk, so that the spore is pushed over laterally. Slight kinks are often visible at the points where such lateral spores have been shed.

Instead of the tip of the conidiophore bearing a single spore or chain of spores, it often branches and the conidia are borne on the ends of the branches, which may be of limited growth and characteristic shape as in Penicillium. Or instead of branching, the apex of the fertile hypha may enlarge into a head or columella as in Eurotium which bears a number of spores or chains of spores, again usually on special short stalks. The term sterigma is again often applied to such a spare-bearing stalk and, as it may branch one or more times before the spores develop, sterigmata of a second or third order may be recognized.

The sterigma, or hyphal branch, which abstracts conidia in succession, thus becoming constricted just below its apes, before cutting off the spore, is often called a phalloid, and the abstracted conidium a philosopher, e.g. Penicillium, Margaritispora, Flagellospora. During this development there is no formation of a cross-wall separating the spore from its phalloid, and a cleavage line arises separating the two when the spore is fully grown.

When, however, a newly developed spore is delimited from its parent hypha by a new transverse wall, it is called an aleuriospore. In some forms it is not certain where the conidiophore and the aleuriospore begins, until the separating cell is differentiated. Again, spores formed on a sterigma or hypha, without reference to the growing point of the hypha, are called radulospores, e.g. Botrytis cinerea. Some fungi develop their spores in two of these ways, thus the last-unatped type may also produce phialospores, and Acrospeira levis may have aleuriospores and phialospores, but others, such as Penicillium produces, asexually, only phialospores. Thallospores, as above stated, are single spores derived from a mere swelling of the end of a hypha, and are not distinct structures, and while the distinction between radulospores and phialosporesis not always sharp, that between thallospores and phialospores is fundamental. Some spores are formed more or less continuous along the hypha, with no sterile internodes in between as are found in sporoaochia, and are called pionnotes in pionnotal cultures sporulation takes place freely within the medium as well as on its surface, but in sporodochial growth, sporulation is almost entirely confined to the surface.

Outside the three main groups, Phycomycetes, Ascomycetes, and Basidiomycetes, there remains a vast assemblage of fungi known only in the conidial stages. These Imperfect species, the Deuteromycetes or Fungi in perfectly are assumed on good grounds to be mainly asexual stages of Ascomycetes.

Their numbers are slightly reduced from time to time by the discovery of the place of one of them in the life-cycle of some as cigerous species, but it seems probable that in many the perfect stage has been lost and the fungus is perpetuated indefinitely in the couidial condition. They include many parasites and amongst them are represented every type of asexual spore and sporophore mentioned above, the simplest forms being designated Hyphomycetes and the pycnidial forms Coelomycetes. Some hesitate to apply the terms genera and species to such imperfect organisms and prefer to qualify them as 'form genera' and 'form species'.

So long, however, as systematics has not reached the point of indicating true affinity there seems to be little reason for this refinement. Names based on characters that can again be identified are necessary to group organisms together into recognizable units, and a fungus capable of independent existence, perhaps indefinitely, in a conidial stage appears to be fully entitled to such dignity of status as may be implied by the use of a properly employed binomial terminology.

Polymorphism

The occurrence of different types of spore and of spore- bearing structures in the life-history of an individual fungus has been indicated already, Ascomycetes, rusts and Fungi being particularly rich in examples. A rust, for instance, may have spermagonia, acedia, uredosori, teleutosori, and the promycelium in its full development. So also an Ascomycete may have a simple Hyphomycetal spore stage and a development of pycnidia, or spermagonia before asci are produced.

In the life-history of some Hyphomycetes the type of spore found may vary, presumably as a result of age and the action of environmental conditions, to such an extent as to place the fungus in different genera according to the prevalent spore form encountered. Thus, the well-known Heteropatella antirrihini, to dispersal, are shown between the cause of 'shot hole' of cultivated snapdragon, may appear as a Pseudodiscosia or a Cercosporella, which is actually only a summer conidial stage of the pycnidial form Heteropatella on the dead or moribund stems: and Ramularia vallisumbrosae the cause of 'white mould' of narcissus, may appear as an Ovularia or a Cercosporella instead of a typical Ramularia. Similarly there is reason to believe that Ovularia primulae and Cercosporella primulae are the same fungus as Ramularia primulae. Amongst pycnidial fungi, again, it is probable that Phoma destructiva and Diplodina lycopersici refer to the same tomato parasite, the perfect stage of which is Didymella lycopersici, while it seems likely that Ascochyta lycopersici is only another phase of this polymorphic species. Species of Septoria and Stagonospora are also stated at times to pass through an Ascochyta stage. It is cases such as these that give force to the arguments against recognising imperfect forms as unqualified genera and species, but many of them are due to such notoriously variable characters as spore septation and sporophore type. Indeed, it is not unknown for variable characters in the sporophore to make it difficult to assign, say, one of the Polyporaceae to its proper genus.

In the opposite direction to the conidial forms mentioned in the last paragraph are those in which similar conidial stages may occur in fungi that are systematically far removed from one another. Thus both the Ascomycete.

The Dissemination of Fungi

A few fungi are disseminated principally in the mycelial state as, for instance, the tropical thread blights which blow about with the leaves to which they adhere; the 'Corticium' disease or 'red thread', a common trouble of lawns and sports greens in Britain, produces on the leaves a mycelium of coral-like threads which adhere to the tips of the dying leaves. Others are distributed as vegetative propagants such as the sclerotia and stromatic crusts of Rhizoctonia solani, of which the sporing form, Corticium solani, plays little part in spreading the fungus. The long-distance dissemination of fungi is, no doubt, often to be traced to the transference of viable mycelium in plant produce from one country to another; this has happened in elm disease and very probably in potato blight and with many other fungi.

Usually, however, the spores are the organs of dissemination. As they are very small and so light as easily to be caught up into the air as an impalable dust, the wind is the chief agent in their dispersal; there are, however, other important methods of spread. The spores of fungi may retain their vitality for considerable periods and they show amazing powers of resistance to extremes of heat and cold: sealed in vacuum tubes, slowly dried spores of Mucor and Aspergillus have withstood a temperature of liquid air for nearly 500 hours, followed by immersion of the tubes in liquid hydrogen for 77 hours. Susceptibility to heat varies greatly, for some spores are killed in 10 to 15 minutes at temperatures far below the boiling point, whereas occasionally some can stand the direct action of boiling water or live steam for the same length of time. So also the spores of some species are killed by drying for 24 hours or even less, while others can be best preserved in a dry condition.

As a rule spores are less sensitive than hyphae to adverse surrounding conditions, and vegetative growth is often inhibited at, or near, the freezing point, as is evident from the action of refrigeration in checking mould growth; slow growth on frozen foodstuffs, however, has been recorded, and very slight superficial moldering of meat will occur at a temperature of – 6° C. Unless special percutaneous are taken, moldering as a result of spore germination quickly supervenes on removal from cold storage. These resistant properties of spores are of value in facilitating the aerial dissemination of many fungi, for desiccation in the air and extremes of beat and cold on the ground may have to be faced.

The vertical elongation of the sporophore is often a powerful aid to exposing the spores freely to air currents. Amongst other adaptations which seem to have the same object, the forcible projection of spores from their stalks is often important. When enclosed in a receptacle there are sometimes arrangements for expelling the spores actively into the air. These are often brought into action by changes in temperature or dryness, as in some Ascomycetes where' puffing' can be induced by concentrating the sun's rays through a pocket~ lens on the.receptacle. Some of the Discomycetes respond similarly if breathed upon when ripe. In the Basidiomycetes it is usual to find a mechanism for the violent release of the basidiospores based on the secretion of a drop of fluid at the junction of the spore and its sterigma. For full advantage to be gained from these various adaptations it is necessary that the spores should be dry and free from mucilaginous envelopment.

In a considerable number of species, including many fungi whose spores are developed in acervuli, the spore beds are surrounded by a gummy secretion. A great many of the bacterial plant pathogens also are immersed in mucilage as they are extruded to the surface of the host plant. When dry, the secretion tends to harden into flakes or sheets, but these are readily soluble in water, as in 'halo blight' of French beans, due to a bacterium.

Dissemination may be effected by the blowing about of the dried gummy shreds, followed by the setting free of the individual cells in rain drops, as sometimes occurs in the American fire blight of fruit trees, due to Bacterium amylovorum, or in Xanthomonas malvacearum in cotton, though the latter is often also carried as dust from adjacent cotton fields infected the previous season. The spore bed may be exposed to rain splashing which carries the spores to leaves or stems in the vicinity; species of Colletotrichum are often disseminated in this way. Amongst the Hyphomycetes it is possible to distinguish two main groups according to the slimy or dry character of the spores, and the method of their dispersal depends largely on this character, the dry spores being predominantly suited for distribution by the wind, while the slimy ones are carried by water, by insects, or by adhering to adjacent plants and the like. In a recent survey of the British Hyphomycetes three- fourths of the species belonged to the dry-spored group. The Coelomycetes, on the other hand, are generally representative of moist sporing fungi. In some of them and in many of the moist-spored Hyphomycetes, the spores are liberated in a mass held together by water or mucilage, and in this form may be subject to aerial dispersal before they are broken up by rain or dew.

Knowledge of the long-distance air-borne dissemination of fungal spores has greatly increased of recent years. Observations on the spread of white pine blister rust in the west of the United States and Canada indicate that it is due to wind-blown infection of currant and gooseberry bushes by aecidiospores from pines that may be hundreds of miles away. Similarly, cereals are infected over great areas by wind-borne uredospores of their rusts. Numerous examinations of the spore-content of the upper air have been made by aeroplane. In early flights, carried out in 1921, spores and pollen grains were found to be fairly abundant up to 11,000 feet, above which they became scarce.

U redospores of Puccinia graminis caught at 7,000 feet were germinated successfully. Above 10,000 feet, mould spores, such as those of Aspergillus and Penicillium, and bacteria, are most frequently found, occurring at times up to nearly 20,000 feet. There is circumstantial evidence that the distance to which such spores may be borne in a viable condition during storms is limited mainly by their ability to survive atmospheric conditions. These naturally vary greatly from year to year during the period of spread, but it is reported from Canada that there is rather conclusive evidence that under certain conditions the black rust of wheat has been carried for four or five hundred miles, or even farther, through the air. These spores tend to be deposited in definite showers rather than in a continuous steady fall; in western Canada rust spore- showers occur after a day or two of strong south wind.

The long-distance disstfmination of cereal rusts has been intensively studied. Black rust due to the well-known parasite Puccinia graminis occupies a practically continuous wheat belt in North America from the Gulf of Mexico to the Prairie Provinces in Canada; as the crop is maturing in the south at the same time that it is being sown in Canada, and the intervening areas are more or less intermediate, a regular succession of wheat ready for infection is available from south to north in the spring and summer. The spores seem to follow a main northward drift path from an area in Texas, where the rust over-winters in the uredo stage and the wheat ripens in the late spring, through Kansas in May and June, to the upper Mississippi States. In Texas itself the summer is intensely hot and the rust is unable to survive until the new crop is sown in the autumn. Reinfection then takes place, apparently chiefly by a reverse flow of spores from the north. So there is an ebb and flow of infection to and from the area in which this fungus over-winters but cannot survive the summer. Confirmatory evidence of this movement is provided by the incidence in particular years of certain physiologic races of the fungus in the areas concerned.

No doubt the spread is not ordinarily carried out in great jumps, though spores not of local origin have been caught as far north as Nebraska in May and June: usually there is time for one or more breaks and the renewal of the inoculum by the infection of local crops in the path from Texas to the northern Mississippi and Canada. Some irregularity of incidence in the upper part of this tract may result from the local over-wintering of the fungus in the teleutospore stage and infection of the alternate host, the barberry, in the spring, for this rust is heteroecious and produces its aecidial stage on barberries, whence it can infect neighbouring wheat and possibly start an epidemic.

But many of the barberries have been destroyed in the barberry eradication campaign and their disappearance in due course will, it is hoped, save much of the chief wheat-growing region in the northern States and Canada from all but the northward spore drift, for the region is much too cold to allow the rust to over-winter in the uredo stage. Attention can then be concentrated on the production of rust-resistant wheats in the comparatively restricted over- wintering area in the south. Nearly all the wheat grown in India is around as a winter crop in the plains.

In these areas rust is unable to survive the intense heat of the summer, and the wheat and barley crops are infected anew each year from rusted cereals grown in the Himalayan foot-hills and some of the hills of Peninsular India. Barberries occur in these hills but. not in the plains, and though the aecidial stage of Puccinia graminis has been found on several of the Himalayan species, it seems to play little part in the perpetuation of the rust. The uredo stage is abundant when the hill crops are harvested in May and June, and then persists on stubble shoots and volunteer plants until the new hill crops are sown in the autumn. Areas have been found in the Central Himalayas where the early crops are sown in August-September and may be heavily rusted by the first week of December. Inoculum is thus available in time to account for the first outbreaks in the Indo-Gangetic Plain, which usually occur in January-February.

The uredospores have been trapped in the air by means of specially devised aeroscope slides we use before the rust has been found on the crops in the plains. In Peninsular India similar foci of early infection of the growing hill crop have been found and seem to account for extensive distribution of the rust in Madras and the Deccan. A similar chain of infection serves to explain the distribution of wheat brown rust, whereas Puccinia glumarum, the cause of yellow rust, is more sensitive to heat and its uredosori are not ordinarily found in the summer below about the 6,000 feet level in the Himalayas, though it may be the only rust seen above 7,000 feet in N ovember- December.

The long-distance dissemination of cereal rusts in several parts of the world is effected on the same broad lines. Thus the wheat in the Amur region of eastern Siberia gets its inoculum from over-wintering areas in Manchuria; North Africa seems to be infected from Europe, the Balkans from Asia Minor, and the Argentine from Brazil. So to a great extent has been cleared up one of the greatest mysteries with which plant pathologists in certain parts of the world were faced before it was known that spores could be carried over long distances in the upper air.

Such long-distance spore flights are evidently limited to certain types of spores and of prevailing meteorological conditions, at least so far as the parasitic fungi are concerned. If they were general, the object of the plant quarantine services maintained by most countries would be defeated. Heavy wind-drifts or storms no doubt play an important part in them, as does the presence in a receptive condition of a susceptible crop in the area reached. It is known, furthermore, that sporecontaminated air not far from the ground purifies itself by the fall of the spores at rates varying from o. 5 to 5 mm. per second, while their presence over the ocean has been found to diminish with increasing distance from the land and catches taken on shipboard may fall to nil in individual voyages fairly soon after reaching the open sea; various favourable conditions must coincide to permit successful long-range spread, and they do not seem often to occur.

There is still a great weight of evidence from the known facts of the spread of parasitic fungi into new areas in support of the view that' discontinuous' dissemination is usually effected by transport on or in the host plant, and is linked with the movement of nursery stock or other plant produce along recognised trade routes. The discovery of spore-passage in the upper air and its bearing on the dissemination of the rusts shows that wind-borne distribution can be an important factor in certain circumstances and must always be regarded as a possible menace by those responsible for the protection of crops against foreign disease.

It is also ordinarily accepted as responsible for much of the' continuous' dispersal of fungi within any given land area. But sea-barriers and sometimes mountain ranges have proved effective checks to the dissemination of many plant parasites. Either the spores are unable to survive long passages in the air, or the timing of their flight is such that a suitable host in a receptive condition is not available when the new area is reached, or the meteorological conditions are adverse, or other causes intervene to spoil a safe lodgement. Many favourable circumstances must combine to render the invasion successful, and experience suggests strongly that they seldom do. Supporting evidence can be found in the history of the spread of American gooseberry mildew in the British Isles and on the continent of Europe; of potato wart disease in many parts of the world, though the resting sporangia of this fungus might be expected to reach the air with dust from infected fields; of blister blight on tea in India; of cacao witches' broom in South America and the West Indies; of several parasites of trees and of ornamental plants, and of many others. In general it may be said that if human agency be excluded, dispersal within a new area is gradual and continuous, each new generation extending the area by a rather limited distance.

This happened when the, vine mildews and that of the oak were introduced into Europe from America, though the latter at least seems to have had some jumps of about 100 miles, and also with several diseases such as chestnut blight introduced into America from abroad.

Some spores are disseminated largely with the seeds of the plants on which they grow. This is the chief means of spreading bunt of wheat and various other diseases against which seed disinfection is practised. Soil fungi are carried by irrigation water or surface-wash after heavy rain and are also liable to be distributed by farm implements, or plough cattle, or on the feet of farm workers.

The dissemination of the spores of certain fungi by insects has been noticed by botanists for very many years. Basidiomycetes belonging to the Phalloid or stinkhorn group were particularly marked examples, for their strong and penetrating odour attracts carrion and other flies, and the special adaptation of this group for insect dissemination was early suggested and subsequently fully established. The sweet honey-dew in which the spores of some fungi are extruded has also long been known to attract insects. These visit the 'Sphacelia' stage of Claviceps purpurea, the cause of ergot of rye, and carry spores away not only as an external contamination but also in the alimentary tract, through which they can pass uninjured. The more recently discovered function of insects in securing the fertilizations of the rusts has been mentioned above.

Naturally, however, it is the role of insects in the spreading of parasitic diseases that has attracted most attention in modern times, and its importance is now fully recognised. Insects are the plain agents of dissemination of virus diseases. They also play a considerable part in the spread of bacterial diseases, such as fireblight in orchards, where they are attracted to the gummy exudate from twig cankers in the spring and carry the bacterium-laden matter to the blossoms and young growth of healthy apple and pear trees; the honey-bee is considered to be one of the chief agents in initiating the blossom blight form of infection, and the discovery that the organism is able at times to persist in the hive was followed by an embargo being placed on honey from infected countries in some places still free from the disease.

Some of the insects concered are able to inoculate healthy trees by puncturing the young growth. The organism that causes bacterial wilt or Stewart's disease of maize, Xanthomonas stewarti, has been found to over-winter in hibernating Chaetocnema pulicaria flea beetles, the chief vectors of the disease in the United States, and to be inoculated into the growing crop by these insects in the spring. Cotton stainers of the genus Dysdercus, and some other insects, are responsible for the introduction of the minute fungi Nematospora and its allies into young cotton bools, where they induce internal boll disease; the insects live in between successive cotton crops on various plants and trees, of which the baobab is one of the most important in Africa, but it is not yet known how or where they harbour the fungus, though the evidence suggests that the association is close.

In many cases insects that cause wounds, such as those that bore into stems or gnaw the softer parts, carry parasitic bacteria or fungi on their bodies and serve to spread infection. The Dutch elm disease is mainly disseminated by bark-boring beetles of the genus Scolytus, and the watermark disease of willows seems to be spread by several different kinds of insects.

The black rot, and soft rot of cruciferous plants, due to Xanthomonas campestris and Bacterium carotovorum respectively, are also carried passively by insects and other small animals such as slugs. It is easy to demonstrate how readily some of these organisms are picked up by allowing an insect from a heavily diseased plant to pass across the solid surface of some suitable nutrient preparation and observing the colonies of micro-organisms that subsequently develop along the insect's track.

The Nutrition of Fungi

The spare, on germination, gives rise to a germ-tube, which at first is nourished by reserve protein, carbohydrate, and fat stared up in advance. In many cases the fat reserves are mare prominent than those of a protein or carbohydrate nature, being often very noticeable as oil drops in the ripe spore. Very soon the young hyphae begin to feed far themselves. The organic food they require must be dissolved before it can be taken in, because solid particles cannot pass through the cell membranes.

This food in solution reaches the interior of the hyphae by a process of absorption. Unlike the green plants, fungi cannot build up their carbohydrate intends by photosynthesis from the carbon dioxide of the air, far they lack the chlorophyll which makes this possible. In this respect the chlorophyll-free plants are like animals requiring their carbohydrates to be ready prepared by having been built up to farm the bodies of other plants or animals. They are usually capable, however, of modifying carbohydrates and other organic compounds by enzyme action sa as to make them mare readily assimilable. In organic substances-mineral salts and elements like nitrogen are as necessary in the nutrition of fungi as in that.of higher plants and animals. Mast.of these, however, occur in combination in the. organic substances.on which fungi feed. As with the higher plants nitrogen, phosphorus, and potassium are amongst the mast important of these elements.others required in lesser degree being sulphur, magnesium, calcium, and a little iron.

It is nowadays a matter of common knowledge that animals and man require certain accessory food substances vitamins and the like in addition to those which satisfy their needs in proteins carbohydrates, fats, and salts. Fungi seem to have similar requirements, though knowledge of these is not yet far advanced. Many fungi and bacteria are able themselves to synthesise these substances and even to produce them in excess and excrete them into the surrounding medium. Various examples of the growth or development promoting action of one fungus an another have been known far a long time, and mast of these can be traced to the liberation of auxins or vitamins from the mycelium of the first of the two organisms concerned. Filtrates of the staled media in which certain bacteria and fungi have been cultivated have the same effect.

Aneurin has been found especially active in this respect. Many fungi are able to manufacture sufficient far their needs or even to produce an excess, but others require an external source of supply. On the whale, this part of the vitamin B complex appears to be the mast generally valuable accessory substance required by fungi.

Yeasts grown in pure synthetic media develop slowly until they synthesise a substance originally named 'bias' but now known in purified farm and commercially available under the name 'biotin' when biotin or various fungal or bacterial filtrates or organic substances containing it are added, growth is considerably stimulated. A good many other fungi are now known to respand by increased growth when biotin is made available to them; same samples of the agar used far solidifying culture media contain it in considerable quantity.

Sometimes two fungi can be grown successfully in mixed culture in synthetic media when either alone makes little progress. Thus, Polyporusadustus and Nematosporagossypii do well together, the former supplying biotin which is essential to N. gossypii and the latter aneurin without which P. adustus virtually fails to grow. Similarly the red yeast Rhodotorula rubra can grow together with Mucor ramannianus in a synthetic medium in which neither alone succeeds. Both biotin and aneurin can be synthesised by certain fungi from pure dextrose, amino-acids and inorganic salts.

According to the manner in which fungi obtain their organic food, they are divided into two great classes: Saprophytes and Parasites. Saprophytes obtain their organic food from the dead tissues of animals or plants, or of substances derived from them. Hence they are found in large numbers in the soil, pervading every bit of rotten leaf or twig or root or every particle of the backs of books in the rainy season. Every bit of old timber in the forests is liable to be permeated all through with their hyphae. As already mentioned, they share with the bacteria the role of nature's scavengers, often having a specified part to play in preparing refractory material, such as the skeletal parts of plants, for bacterial decomposition. In breaking down the walls and contents of plant cells, they liberate carbon dioxide which becomes available in the air for the carbohydrate nutrition of green plants, hydrogen which forms water, ammonia, and so on.

Parasites obtain their food from the living tissues of animals and plants. While, therefore, many saprophytic fungi playa useful part in nature as scavengers, the parasites are almost entirely hurtful. The few cases of beneficial parasites, from the point of view of man, are those which attack noxious insects and occasionally destroy large numbers of them. The great majority of parasitic fungi feed on living plants, those useful to man as much as, or more than, the weeds or useless kinds. A living plant on which a fungus feeds is often termed a 'host plant' for that fungus. There is scarcely a plant of our fields, gardens, and forests which does not serve as a host plant for one or more species of fungi.

Several different classes of parasites may be distinguished. Obligate parasites, for instance, pass through the whole of their life-history on living plants, and cannot be grown on dead or artificial food material. Such are the rusts, the powdery mildews, some Ascomycetes, and other fungi. Facultative saprophytes are those that feed usually on living tissues but can, at need, pass through a part of their lives as saprophytes. Often their full development is only reached on the living plantsaild they cannot be got by artificial cultivation to complete the whole of their life-cycle.

The smuts furnish examples of this, as they can often live by budding out little conidia for a considerable time in soil, but usually only grow to maturity and produce their perfect spores on a living plant. Facultative parasites, on the other hand, are those which usually grow on dead or decaying matter and are capable of passing through the whole of their development as saprophytes, but have at the same time the faculty of attacking living tissues under certain conditions.

Within these main divisions, parasitic fungi may show every degree of adaptation to the parasitic mode of nutrition. At the lowest end of the scale are certain facultative parasites, which may be called' weak parasites', as they become parasitic only occasionally, when the host plant has been weakened in its vitality by some harmful agency. Plants not yet acclimatized in a new locality, or suffering from insect attack or mineral deficiency, or grown in insanitary conditions such as overcrowding, too dense shade, unsuitable or badly drained soil, and the like, are particularly liable to become available as food to these fungi that are normally saprophytic. Cases of the sort have been recorded from time to time, though few have been adequately investigated; for instance, Diapor the perniciosa and other fungi to which the dieback and bark canker of fruit trees in the south of England were at one time attributed are now believed to attack only trees weakened from other causes. It is, perhaps, as profitless as it is impossible to draw a hardand-fast line between saprophytism and parasitism in some of the fungi; even in the commonest of all species, Cladosporium herbarum on fading green plant parts, there are too many records of injury from its action to deny it all possibility of behaving as a weak parasite under certain conditions.

In many cases, particular tissues of a plant are normally of low vitality and are readily available as food for weak parasites. Such are the gorged cells of many pulpy fruits, oranges, cherries, apples, and the like, and also the older living wood of trees.

It is probable that the fungi termed 'wound parasites' are limited in their aggressiveness by some nutritional failing. At first these live only on the dead and rotting tissues caused by a wound or injury to the plant. Later on they begin to spread to the undamaged cells farther in, and they may extend through a large area of the living tissues, eventually, perhaps, killing the whole plant. It may be that during the period of saprophytic life the enzymes of the fungus are produced in quantity and concentration sufficient to enable them to attack living tissues exposed to their action, and transform the walls and cell-contents into nutrient material which the fungus can utilise. If this is so, an analogy can be drawn with the action of decaying petals in promoting infection by Botrytis cinerea. That enzyme production can be greatly modified by nutrition has been fully established in this and other fungi.

The distinction between parasites and saprophytes cannot be maintained in a few fungi which are able to attack perfectly sound tissues of healthy plants should they come in contact with a suitable host, but in the absence of the latter can live and develop freely on dead materials. Such a case is the' damping off fungus Pythium de baryanum which is present as a saprophyte in almost every garden in some parts of Europe, and at once attacks and destroys seedlings of cress and certain other plants, should they be grown in the soil which contains the fungus.

Neither neglect of cultivation nor unsuitability of soil predisposes to this disease; wounds are not necessary for the entry of the parasite; and, if anything, damping-off is worse on rich soil than elsewhere. Several parasites of trees, e.g. the common bracket fungus on elms, Polyporus squamosus can continue to live on the dead wood as saprophytes; these, however, are usually wound parasites of a less aggressive type than the damping-off fungi. On the Aleppopine in France, Trametes pini dies with the tree, but Fomes pinicola continues to rot the fallen timber.

The Mechanism of Feeding by Parasites

Parasites reach their food either by sending their mycelium into the interior of the host plant or by remaining for the most part superficial and sending only suckers or a limited growth of feeding hyphae into the surface tissues. Usually in the internal parasites, e.g. Rhytisma acerinum, the feeding hyphae penetrate directly into the cells of the host, the cell sap and more solid contents being taken up by absorption after the latter have been brought into solution by the enzymes of the fungus. Often, however, the earlier stages of feeding are marked by the dissolution of the cell walls of the host by enzymes such as pectinase and cellulase; these walls can provide many fungi with much of the carbohydrate food that they require. In a few fungi, such as the Exoascaceae, the mycelium is wholly intercellular, without any intracellular parts, so that absorption of nutrient solutes must occur through the membranes of the host and parasite cells; this is probably a common occurrence, for even when the cells are entered much of the mycelium may remain intercellular and probably receives its food directly from the surrounding cells. It was at one time believed that some of the superficial parasites of leaves and twigs could similarly feed by absorption from the epidermal cells across the outer wall of the cell and the cuticle, but in most of these fungi, feeding hyphae or suckers have now been found to enter the cell 'cavities; this occurs, for instance, in the light- coloured thread blights and in the dark Meliolaceae or Asterineae.

In the ectoparasite of some European pine trees, Acanthostigma parasiticum, it is stated that the white mycelium which forms cushion-like layers on the under surface of the needles obtains its food through tiny rod-shaped processes which penetrate through only the cuticle and not through the inner parts of the outer wall of the epidermal cells. As in the subcuticular parasites such as Taphrina rhomboidalis, absorption of nutrient solutes must occur through the subcuticular membrane of the needle leaf.

HAUSTORIA

Many intercellular parasites that apparently find difficulty in getting adequate nutrition by absorption of solutes into their hyphae from the cavities of the host cells send into the cells specialised hyphae which act as feeding organs. These are known as 'haustoria'. Haustoria occur in all the main groups of fungi, usually but not exclusively in obligate parasites with an ectoparasitic or an intercellular mycelium. They vary in shape from the knob or saclike type found in Cystopus to simple or lobed or branched or coiled hyphae in many Erysiphaceae Phycomycetes and rusts, or coralloid types as in Peronospora schachtii on beet. Some of the complex forms almost fill the cells they occupy and can hardly be distinguished from an intracellular mycelium, as in Peronospora parasitica. Their shape is not very constant, as it may vary to some extent according to the tissue invaded.

The haustorium usually enters the cell by an exceedingly fine hypha resembling the infection hypha arising from an appressorium by means of which, as will be seen later, many parasites enter the outer membrane of the host. After entry the haustorium often grows towards the nucleus of the host cell but while, at first, it may not enter the' primordial utricle' or peripheral plasmatic membrane of the protoplast, only pushing it in and invaginating it, the utricle may finally be pierced by it. In the early stages the haustorium is bounded by a thin limiting membrane, and this sometimes persists. Frequently, however, various thickenings occur around the point of entry or deeper in.

When developed around the narrow style-like stalk of the haustoria of Erysiphe graminis, or many rusts, the thickening is continuous with the inner wall of the cell and is closely similar to the infection pegs found where a parasitic hypha enters a cell. It is sometimes effective in checking penetration. When found around the body of the haustorium it is often later in appearing and may, in old haustoria, be a means of walling off the intruder, or a consequence of excessive drain on the cell contents, as in tubers attacked by potato blight; as injury to the cell from over-virulent action of the fungus increases, so the haustorial cover thickens.

There is little doubt that these thickenings are products of the host cell, and indeed it is sometimes possible to distinguish two layers in the sheath, one being the membrane of the haustorium and the other a covering deposited by the host. The latter seems to be usually cellulose, but there may be some lignification of the older sheaths. Callose is stated to occur sometimes between the two layers or in the exterior one. The haustorium itself may have from one to many nuclei. In some smuts the haustoria are typically binucleate, but those of the downy mildews may sometimes have only one nucleus though others have several, as in the coenocytic hyphae from which they arise.

Most of the haustorial fungi arc obligate parasites, but some, such as the Phytophthoras and Actinonema, can be grown in artificial culture. It is probable that many of the intercellular hyphae are not wholly dependent for their nutrition on the haustorial apparatus, but, like other fungi occupying a similar position, can get some nutrition by absorption from the surrounding cells.

While, therefore, the haustorial habit is not a strict criterion of status as an obligate parasite, it suggests a high degree of specialisation in some requirement which ordinarily it has not proved possible to supply in culture. It marks increased adaptation to the parasitic life over that shown by the destructive action of parasites such as Pythium and Botrytis, for one of the characteristic features of many of the haustorial fungi, especially the rusts and smuts, is the absence of any gross indication of injury to the protoplast of the host cell for a period which may sometimes be almost equal to the lifetime of the plant. This aspect of their parasitism will be referred to again in a later chapter.

The Fungi Enzymes

In the preparation of their food for absorption fungi, like animals and man, rely on enzymes or ferments, of which they may possess a formidable array. Among the hydrolytic enzymes detected in Polystictus abietinus are cellulase, pectinase, ligninase, diastase, sucrase, inulase, tanninase, pepsin, trypsin, emulsin, and lipase, while the oxidising enzymes of this fungus include catalase, peroxidase, and laccase. In Aspergillus oryzae diastase, invertase, maltase, lactase, protease, rennet, and lecithinase have also been recorded.

Many of these enzymes are excreted into the surrounding medium, sometimes in considerable quantity and capable of utilisation on a commercial scale. Apart from the use of yeasts in brewing, distilling and baking, diastase from fungi is used medicinally and in various industrial processes; acetone can be formed by Mucor rouxii and other fungi from sugar-yielding plant products; citric acid is now mainly a fermentation product of Aspergillus niger, for which one American firm some years ago kept nine acres of mycelium in constant commission; yeasts, Mucors, Aspergilli, and bacteria are all employed singly or in various combinations in the East and in Africa for the preparation of various 'beers', fermented milks, non-alcoholic drinks, and seasonings; while gallic acid, which is used in dyeing and as a constituent of inks, has been commercially produced by the fermentation of tannins. The proteolytic enzymes of Aspergillus oryzae and its allies are used in the Far East in the preparation of readily digestible substances from the protein-rich soy bean.

Not less important are the processes of fermentation carried out during metabolism within the hyphae. During the period of stringency of fats in Germany about 1918 more than a million kilograms of glycerine were being produced monthly from fungi, and in Russia Endomycopsis vernalis was used experimentally for the same purpose some years later. Molasses, beer-wort, potatoes, hydrolysed wood pulp, or the like, together with salts of nitrogen and phosphorus, provide the culture medium for the fungus. Some moulds are even more effective in synthesising fat as a reserve nutrient in their mycelium than the yeast-like Endomycopsis; Penicillium javanicum can store up to 41.5 per cent of fat in its hyphae when grown on glucose and salts.

Yeasts and their allies are also able to manufacture high-quality proteins, with vitamins, from the same raw materials, and some 20,000 tons of dried yeast, containing about half its weight of protein, were annually produced in Germany during the first great war. Probably the best species for this purpose hitherto tested is Torula utilize. These examples, taken from only a few amongst many, indicate that both by means of the enzymes excreted from the mycelium and by those concerned in metabolic processes within the hyphae fungi can synthesise a remarkably wide range of substances from organic materials. Many have been identified and they represent almost every type of compound known to organic chemistry.

The enzymes are organic catalysts which, unlike the inorganic ones, are destroyed by heating. Hydrolysing enzymes split up disaccharides such as sucrose, maltose, lactose, and the like into two similar or dissimilar molecules of monosaccharides-dextrose, levulose, mannose, galactose etc. Cane sugar, for instance, is readily' inverted' by invertase from fungi giving one molecule of glucose and one of levulose, while maltase gives two molecules of glucose from maltose. The hexose monosaccharides, especially glucose, appear to be readily available to fungi; on condensation they give hexosan polysaccharides.

The pentose monosaccharides, such as xylose and arabinose, condense with other sugars to form pentosans- polysaccharides common in pectic membranes, gums, mucilages, and the like; acted on by the pectinase of fungi, they give galactose and other bodies. On splitting up, the polysaccharides give more than two molecules of monosaccharides. The proteolytic enzymes of fungi, pepsin, trypsin, erepsin, etc., change the less soluble proteins into the more soluble and readily diffusible peptones and amino acids. Tannase forms glucose from tannin and lipase, glycerol and higher fatty acids from fats. Certain moulds can grow in the waste liquor from the manufacture of sulphate wood pulp, decomposing the lignin present as chemically combined lignosulphonates.

The oxidative processes of certain bacteria provide energy enabling them to synthesise carbohydrates when carbon dioxide is their sole source of carbon, although they have no chlorophyll and light is not needed for the purpose. Such are the nitrifying bacteria, Nitrosomonas and Nitrococcus, the sulphur bacteria. Beggiatoa and Thiothrix, and the iron bacillus Spirophyllum. The oxidase of strains of the common mould Aspergillus niger, can oxidise glucose to gluconic acid directly. Similarly some yeasts can directly oxidise maltose, giving off carbon dioxide. Yeasts will grow with the organic salt ammonium tartrate as their only source of carbon. Such examples are sufficient to illustrate the remarkable versatility of the enzymes of bacteria and fungi.

The enzyme armament varies considerably in different fungi. Some can attack chiefly the cell membranes, others the cell contents. The Peronosporaceae, Erysiphaceae, and rusts, for instance, have little or no action on the membranes, using chiefly some of the cell contents, while those wood-decaying fungi that affect tissues with scanty cell contents must get much of their food from the walls; this accounts for the cavities in the older wood caused by certain species. Enzyme production may vary according to the strain of the fungus concerned, as is fully recognised in brewing and other fermentation industries.

The conditions under which the fungus grows, especially its nutrition and aeration, also have a marked effect. Thus a strain of Aspergillus niger was found to form diastase copiously when grown in media containing starch, glucose, or maltose but not at all when grown on 5 per cent glycerine. An acid reaction was necessary for optimum production on the enzyme in this case, but pectinase may sometimes be most active in acid, sometimes in alkaline plant juices, and this seems to depend on the presence of other substances which are adsorbed from the medium.

It is believed that the pectinase is the same in all of a number of fungi tested, but its properties depend on its exposure to the action of other substances, while whether it is produced at all and the quantity produced, depend on the medium in which the fungus is grown. Some fungi can cause notable rises in temperature in decomposing plant material. Thus Aspergillus fumigatus has been found to raise the temperature of oat straw inoculated with it from 25° to nearly 55° C., in 38 hours. The most extensive decomposition of straw has been obtained by a mixture of soil Actinomycetes and fungi. Most of the common soil moulds Aspergillus, Penicillium, Trichoderma, Rhizopus, and so forth utilise the carbohydrates cellulose, hemicellulose, pentosans, and mannan readily and rapidly in decomposing plant material but not lignin; certain Basidiomycetes! however, such as the common edible mushroom, do not make use of hemicelluloses but deplete the lignin and protein, while causing also a loss of cellulose. Some of these fungi can decompose humus.

Nutrition and Life-history of Fungi

All saprophytes and many parasites can be grown on artificially prepared food, or 'culture media'. The kind of food supplied may have a profound effect on the structure and reproduction of the fungus. With such plastic organisms, however, generalisations are difficult and the most that can be said is that a particular effect is obtained with a given strain of the fungus grown under a certain set of conditions. Other strains and other conditions may give different results.

In some fungi, as in higher plants, it is possible to enhance vegetative vigour at the expense of reproductive activity by nutritional treatment. Thus, the aquatic fungus Saprolegnia mixta tends to produce a vegetative mycelium only, without any spores, when supplied with abundant food regularly renewed. If such sterile, strongly growing cultures be transferred to water, or to a weaker food solution, sporangia with zoospores are formed in quantity.

Transferred to a solid substance of little nutritive value, such as plain agar, sexual oospores are alone produced. In early studies of Pythium ultimum only oospores were produced on cabbage leaves whereas on other substances sporangia were formed. The mycelium of various common moulds is quite different in structure with different food supplies; in some of the Mucors it may be changed wholly or in part from the filamentous to the gemmate condition. Nutrition may determine the proportion of aerial to submerged mycelium or the formation of coremia in place of single hyphae, or even of sporodochial erect clustered hyphae instead of a recumbent growth.

In some genera the colour of the mycelium may vary according to the composition of the medium on which it grows and also according to the pH of the latter. Such effects are the commonplace of all mycological laboratories and account for the caution shown in attempting to identify fungi from mycelial characters alone. As already mentioned, the morphology of the spores and their method of attachment to the sporophore are much more reliable specific characters and are those mostly used. Even here, however, there are difficulties, linked largely with vigour or luxuriance of development. To surmount them, many futile attempts have been made to devise a standard or an ideal culture medium on which fungi would give consistently satisfactory growths.

That these must fail in, at least, the considerable number of facultative parasites which show the specialisation of parasitism described in the next chapter is evident, if specialisation is dependent on nutritional qualities of the host, and there is overwhelming experience available to show that even pure saprophytes do not all use the same food and cannot all be given a single satisfactory diet.

Symbiosis

The parasites mentioned so far obtain their food at the expense of the host plant and are not known to give anything in return. The food-supplying cells are usually killed eventually by the excessive demands of the invader, but, as already mentioned, certain obligate parasites, especially those that feed by haustoria and those that form galls, may long defer this critical period and kill only when the production of spores makes a culminating call on the nutritional resources of the host. In many cereal and grass smuts this does not happen until the ear or individual caryopses are developing at the end of the vegetative life of the host. The Ascomycete Epichloe typhina may go to a stage further than this, for it sometimes fails to fructify even when its systemically infected grass host has grown to maturity.

The mycelium may be found in the anthers and around the embryo of the grass seed without necessarily destroying the pollen or preventing the seed from germinating. Latent infection may persist for several seasons, the plants sometimes remaining barren, sometimes producing viable seed. Ultimately the conidial stromata appear during a flowering period, which may be in the second or third year of growth; up to this time no apparent injury is suffered by the host. Perhaps the end point in this progress towards a regulated parasitism is marked by the endophytes which occur at times in various species of Lolium. Two of these have been found to cause systemic infection of the above-ground parts of the hosts, one difficult to isolate and remaining sterile in artificial media while the second grows readily in culture and produces microconidia: the former develops a dense mycelial stratum between the seed coat and the aleurone layer, the latter is sparser and more difficult to detect.

Both are transmitted in the seed, from which they pass to the growing point of the seedling.

Neither appears to cause any injury, and their presence in or absence from a particular plant cannot be detected without isolation or microscopical methods. Their identity is unknown, and there is no real clue to the physiological relationships between the host and the parasite, though there is some experimental evidence which has been considered to indicate that the endophyte assists in the supply of nitrogen to the host by fixing the nitrogen of the air.

So far all the advantages seem to lie with the parasites; the host at best may be able to escape injury to a greater or lesser degree, or perhaps, in Lolium, gain a little nitrogenous food by the aid of the fungus. Some higher plants, however, have progressed beyond this stage and have established a symbiotic partnership with fungi in which a mutualistic relationship is evident. As the association between the fungus and higher plant is chiefly evident in roots, the term 'mycorrhiza' is applied to it.

The best-known mycorrhizas are those of forest trees and orchids: these represent the two main types of fungus-root association, the 'ectotrophic' and the 'endotrophic'. In the ectotrophic mycorrhizas of trees many of the fine lateral rootlets are enveloped in a mantle of fungal tissue which prevents growth in length of the rootlet, causing it to assume a swollen or coralloid form. Some of the hyphae usually pass down between the cells of the outer layers of the root, forming the so-called Hartig net.

The fungi concerned chiefly belong to the Basidiomycetes and include many gill and pore bearing toadstools but truffles grow in mycorrbizal association with oaks and other trees and the systematic position of some mycorrhiza-forming fungi is unknown. Field mycologists are well aware of the common occurrence of particular species of fungi under certain trees of which this is, perhaps, one explanation. The mycorrhizal association in the orchids differs widely from that of trees. The fungi concerned are endophytes, mostly sterile Rhizoctonia stages of the genus Corticium, though the honey agaric, Armillaria mellea, forms mycorrhiza with the tuberous orchid Gastrodiaelata, in the East. In the Rhizoctonia type, the hyphae enter the cells of the cortex of the root, where they form dense coils which eventually break down into an amorphous mass. It has long been known that orchid seeds are difficult to germinate, and the surmounting of this difficulty by providing the seeds with cultures of a suitable strain of Rhizoctonia is a good example of the application of science to practice by many orchid specialists, who not infrequently have kept their methods as a closely guarded secret.

The heaths, in general, resemble the orchids, except that the endophytes, belong to the genus Phoma and in some cases e.g. in the common ling, Callunavul occur as a sparse mycelium through out the plant as far as the fruit capsules, whence the seed on germination may become naturally infected. In these two families and a few others similarly dependent on fungal infection for satisfactory germination, the seed is small and lacking in food reserves, so that it would appear that the endophytes supply some nutritional requirement. What this is not definitely known, but the fact that satisfactory germination of orchid seeds occurs without the fungus in certain sugar solutions suggests that carbohydrates are involved. In the heaths a fixation of atmospheric nitrogen has been attributed to the endophyte, but the evidence for this has been questioned.

There is much difficulty in accounting for the physiological relationships between the partners in a mycorrhizal association. Several converging lines of evidence are strongly indicative that it began as an attack by a parasite on the roots of a higher plant the prevalence of infection pegs and other reactions of the host, the occasional tendency for the fungus to develop destructive capabilities, the peculiar features of the mycorrhizal function of Armillaria mellea, are some of these. The eventual reduction of the invading hyphae to a condition in which they are not only made harmless but actually in large part digested by the host cells is easily reconcilable with the view that the higher of the two partners gains nutritional benefits in the process. But even this view, which is far from easy to prove, has been the subject of adverse criticism, with the result that, so far at least as concerns forest trees and orchids, the practical man was for a time more universally ready to accept it than his scientific fellows. Gradually, however, the beneficial effects of mycorrhiza have gained almost general recognition, though there is still a lack of agreement as to their nature.

The fungal partner in endotrophic mycorrhiza gains food, often using up all the starch in invaded cells, but beyond escaping the competition of ordinary soil saprophytes, seems to profit little from its colonisation of the roots. The ectotrophic forms seem to have even less chance of benefiting from the association, though it may be presumed that those of their hyphae that grow down between the host cells can absorb nutrient material from the host. There is some evidence that Boletus elegans gets a growth-promoting substance from the larch roots with which it forms mycorrhiza. Aneurin stimulates the growth of several mycorrhizal fungi.

An entirely different type of mycorrhizal association is found in the majority of annual and perennial plants, in which the cortex of the root is invaded, usually in limited areas behind the apical meristematic zone, by a sterile fungus characterised by special organs known as 'vesicles' and' arbuscules. The vesicles are large thick-walled diverticula or terminal swellings on an inter- or intracellular mycelium, while the arbuscules are haustoria- like, finely branched bodies formed often in particular cortical cell layers. Fat is stored in quantity in the vesicles and larger hyphae and is liberated into the host cells through the arbuscules, which collapse into formless clumps of indigestible fungal elements. The fugi concerned are regarded as probably members of the Zygomycete family Endogonaceae and have been assigned to an imperfect genus, Rhizophagus, of this family. It has been suggested that they may supply, with the fat, some accessory nutrient or growth promoting substance to the host. The last-mentioned type of mycorrhiza is often casual in its formation and there is little evidence that it meets an obligate need of the host. But, in general, there is little doubt that mycorrhizas are beneficial and in certain cases are of great value in securing a satisfactory growth.

The lichens form the most prominent case of symbiosis in which fungi are concerned, though of lesser interest to the plant pathologist than those mentioned above. They consist of a fungus and an alga living in such perfect association that they appear to form quite definite plants. The fungus hyphae surround and grow between the algal cells, into which they sometimes send haustoria; they feed in part on the organic material which the alga builds up by photosynthesis from the carbon dioxide of the air.

The fungus in return appears to supply water, minerals and possibly peptone; it seems to have the best of matters, for it alone forms spores in most cases. The best-known symbiotic association in which bacteria take part is that of the root nodule organism of the Leguminosae, Rhizobium or Pseudomonas radicicola.

This is of considerable importance in agriculture, for many pulse crops will not grow well unless furnished with a suitable strain of the bacillus. The latter has the power of fixing atmospheric nitrogen and making it available to the higher plant and also indirectly to the soil when the crop decays or is ploughed under as green manure. Strains vary in their power to do this; some have been found in Great Britain that are poor nodule formers yet can outgrow and overcome the better types. It may well be that the association started in this case also as a parasitic attack by the bacillus on the root, for under certain conditions Pseudomonas radicicola can become parasitic. The bacillus profits by the same freedom from competition by soil saprophytes that is enjoyed by root parasites. In advanced stages of the disease dark concentric rings are visible in the flesh of the sugar beet when it is cross sectioned.

In tomato, leaflets of infected plants roll upward along the midrib and become thickened and crisp as the leaf curves downward. The leaf tissue gradually turns yellow, while the veins take on a purple tinge. The plant assumes an erect habit and remains stunted. In squash young leaves have a savoyed surface or an upward roll; internodes remain short; many blossoms fail to set; and the tips of runners turn upward. Plants infected when very young are killed rapidly without the above symptoms in warm weather.

Older vines decline gradually and die prematurely. In bean, young affected plants may show killing of the growing point and yellowing of the primary leaf, followed by death of the plant. When older plants are infected, there is a downward marginal rolling of trifoliate leaflets, followed by brittlenees, dwarfing, and puckering. The whole plant may become stunted and unproductive.

The Virus

- Beet-curly-top virus.
- Sugar-beet virus 1.
- Beta virus 1 K. M. Sm.
- Chlorogenus eutetticola Holmes.
- Ruga verrucosans Carsner and Bennett.

The physical properties of the virus as determined by Severin and. Freitag and Bennett are inactivation in vitro in extracted juice, more than 7 days; by dilution, above 1-20,000; by heating, 75 to 80°C. "The virus is unusually resistant to alcohol and acetone and survives in solutions up to pH 9.1. In the phloem exudate, it withstands freezirig up to 18 months. It passes _commonly used porcelain filters. It remained active in thoroughly dried young sugar- beet plants for 8 years, leaf hoppers for 6 months, and in dried phloem exudate for to months.

DISEASE CYCLE

The virus is perpetuated between crop seasons in wild hosts and it may overwinter in the leaf hopper. In semiarid regions such as southern Idaho, abandonment of land and overgrazing often tend to encourage plants which are good overwintering hosts for the insect and which may serve as reservoirs for the virus as well. The injury to beets by transmission of the virus by the insect is dependent primarily upon the time and magnitude of spring leaf hop per movement-into cultivated fields.

Factors which determine the magnitude are size of fall migration to winter feeding areas, winter survival, and spring reproduction; these in turn are influenced by the type and amount of fall, winter, and spring host plants. The date of the spring migration to beet fields is influenced by the prevailing temperature. The direction of the migration is determined by prevailing winds, which are usually from the west.

The only vector known is the beet leafhopper; Eutettix tenellus Baker. The extract from diseased plants or infective hoppers may be used as a source of inoculum, experimentally, by allowing hoppers to feed upon it and by transferring them to the host plant.

The virus has not been transmitted mechanically. Early investigators found that hoppers were not infective until 1 to 44 days after feeding on diseased plants. Sever in secured transmission in a small percentage of cases within 20 mint. Bennetit and Wallace were not successful in securing transmission in less than 4 hr., although the insect picked up the virus within 1 min. Freitag and Bennett and Wallace concluded that the virus does not multiply within the insect.

Carsner secured an attenuation of the virus by passage through Chenopodium murale L. Lackey showed that virulence could be restored by passing the attenuated virus through Stellaria media Cyr. Giddings isolated 10 distinct strains of the virus. He also presented evidence to show that infection was defendant upon mass action of the virus introduced.

The relation of the virus to the host tissue and the reaction of the latter to infection has been studied critically by Esau. The virus in leaves apparently is contained entirely in the phloem during the time it increases and is transported. After one or more of the primary sieve tubes have differentiated, the first histological signs appear in the form of hypertrophy and death of the pedderm and phloem cells adjacent to the sieve tubes. In cells somewhat farther removed from the sieve tubes, hyperplasia occurs; the cells involved go through some of the changes characteristic of differentiation, of sieve tubes, but normal maturation is not complete. These primary phases of hypertrophy, necrosis, and hyperplasia are followed by secondary necrosis in the abnormal sieve tubes aped their companion cells.

The parenchyma cells around the centres of secondary necrosis next undergo secondary hypertrophy and hyperplasia, resulting in preparation resembling wound periderm reactions. Hypertrophy and hyperplasia are not Confined to the phloem. The mesophyll cells adjacent to the phloem are stimulated to grow and divide so that they become closely packed together without intercellular spaces, while the chloroplasts show signs of degeneration. This internal change in the tissue is responsible for the macroscopical veinclearing noted as an early symptom. Further hypertrophy and hyperplasia in cortical tissue result information in leaf and petiole tissue, including the formation of the characteristic protuberance. Lackey demonstrated the virus, in the ineristematic tissue of the root tips.

Bennett showed that if the virus is introduced into the distal end of a beet leaf in the dark, it often fails to move out of the leaf in a period of several days, whereas when it is inoculated in a green leaf in the light, it moves out in as short a period as 4 hr. He believed that the movement of the virus has little relation to its multiplication or to its concentration but is correlated with food transport and takes place entirely in the phloem.

VARIETAL RESISTANCE

As already indicated, numerous varieties of sugar beet resistant in various degrees to curly top have been developed and released for use in various areas to which they are adapted. All such varieties are carriers of the virus and may show signs of the disease under optimum conditions.. They are genuinely tolerant, however, and the amount of phloem necrosis is greatly reduced, as shown by Bennett and Esau, while the actual amount of virus present in the tissue is less, as shown by Giddings. Lackey found that in a fairly resistant Great Northern bean the virus was promptly inactivated in the root tip, while it persisted in the susceptible Bountiful variety. Inheritance of resistance in bean has been studied by Schultz and Dean.

Control

Band and crop management and controlled grazing are important measures in reducing as much as possible the magnitude of the spring migration of leaf hopper from overwintering areas to cultivated crops. Early planting and liberal fertilization of resistant varieties of sugar beet are desirable. At present certain-susceptible varieties and species must be avoided in areas where the chances of heavy infestation by viruliferous hoppers are great.

Beginning with the resistant variety of sugar beet U.S. 1, there has been developed a succession of varieties of greater agronomic value and adaptability to various areas. In southern areas, where culture comes during winter months, easy-bolting varieties must be avoided, although such a variety may be quite satisfactory for a summer crop in northern areas. The variety U.S.,22 occupied most of the Idaho acreage in 1942, and an improvement of this line was released in 1946 as U.S. 35. In the same year a new non-bolting curly-top-and downy- mildew-resistant variety U.S. 36, was released-Jar use in the San Joaquin valley and in the Imperial valley of California.

Dana apointed out that Umatella Marblehead and Yakima Marble head varieties of squash are spaturally highly olerant to curly top and are of particular value in the Pacific Northwest.

In southern Idaho the Common Red Mexican bean is resistant to curly top, while the more widely grown strains of the variety Great Northern are susceptible to curly top but resistant to common mosaic. From hybrids between these two parents a new variety of the Great Northern type resistant to both virus diseases, has been introduced as Great Northen U.I. 15. Pioneer is a curly-top-resistant snap bean introduced by the U.S. Department of Agriculture and the Oregon Experiment Station.

Spotted Wilt

The disease known as spotted wilt was first defined as such in 1919 on tomato in South Australia, where by 1925 it had become a serious disease causing enormous losses. It has since been found oil many crop plants and wild species in continental United States, Hawaii, Great Britain, the European continent, South Africa, Brazil, Argentina, and New Zealand. Among the economic host plants besides tomato are tobacco, potato, pepper, lettuce, endive, celery, spinach, pea, and numerous ornamental species including sweet pea zinnia, nasturtium, poppy, dahlia, calla lily, orchid, chrysanthemum, petunia, aster, schizanthus, salpiglossis, cineraria, and gloxinia. The yellow spot of pineaplle destructive in, Hawaii was shown to be incited by the spotted- wilt virus. A list of hosts was compiled by Smith.

The virus concerned is distinative in that it is transmitted by species of thrips, rather than by one of the more common groups of vectors of plantinfecting viruses.

SYMPTOMS

Since the disease is most widely known and commonly destructive on tomato, the symptoms on that host have been studied most extensively. In all hosts, it is well to not the symptoms may be mild under certain conditions and quite different, symptomatically and more severe in another environment. On tomato one of the first characteristic signs is bronze-coloured marking on the upper side of young leaflets, which may be accompanied by a slight downward curling of the leaves. The intensity of the bronzing varies, and it is undoubtedly influenced by the environment; it may extend to petiole, stem, pedicel, and calyx. Growth may be checked permanently or only temporarily. Upward marginal rolling and stiffening of leaflets follow, and small circular necrotic spots commonly appear on the leaves under optimum consolations. Necrosis extends to the stem near the growing tip, causing the latter to wilt and die Some mottle, chlorotic spotting, and malformation may occur on leaves. On the green fruits, yellowish spots up to 1 em. in diametre appear, often with distinct concentric zones 01 shades of yellow or bronze alternating with green and later with pink or red. The zoned fruit spots are probably the most characteristic sign of spotted wilt on tomato.

On tobacco circular necrotic spots of various sizes appear on young leaves. A pronounced mottle may follow, often confined largely to the lower half of the older leaves, while necrosis of the leaf veins adds to the irregularity of the mottle pattern. Plants are stunted, and leaves may droop gradually and die. Leaves which are mechanically show local lesions up to 4 or 5 mill. in diametre.

Symptoms vary on the numerous host plants, and often times symptoms are so variable as to be of little diagnostic value. Various types of zoned chlorotic and necrotic spots run through the sympto. ❐

8.

Plant-parasitic Nematodes and their Management

The name "nematode" comes from the Greek words: nema, which means "thread", and toid, which means "form". Nematodes are usually vermiform, long and slender, but some species are swollen. Most people know them as roundworms because their cross-section is round. They range in size from .08mm to 7m. Nematodes are a tube within a tube. The outside tube is the body wall which consisits of muscle layers that are used as a protective covering. The inside tube is the digestive system. The esophagus is the front of the digestive tube.

FEATURES OF NEMATODES

Aquatic, triploblastic, unsegmented, bilaterally symmetrical, roundworms, that are transparent, colourless, usually bisexual, and worm- shaped (vermiform), although some can become swollen (pyroform). Only an estimated 3 per cent of all nematode species have been studied.

HOME OF NEMATODS

Nematodes exist almost everywhere, including soil, plant and animal surfaces and interiors, decomposing life forms, and fresh and salt water and their subtending sediments. Nematodes can be grouped according to what they eat. The different groups are fungal feeders, bacterial feeders, predators, animal parasites, algal feeders,omnivores, and plant parasites.

MODE OF INFECTION

Plant-parasitic nematodes pierce plant cell walls with their stylet, which is formed by some of the mouth and esophagus parts. Then they pump up the plant cell just into their digestive system. These nematodes only feed on a feed only on the outer tissue of the root, others penetrate more deeply, and some completely enter the host. There are different ways they harm the plants. Some induce their hosts to produce nutrients which the nematode can survive on, or enlarged structures in which the nematodes live, or both. Some produce metabolites which kill host tissue.

These activities all result in energy being removed from plants to support nematodes. Some plant-parasitic nematodes can cause severe damage that can result in unmarketablity, yield decrease, or even total crop failure.

SOME PLANT-PARASITIC NEMATODES ARE HARMFUL TO OREGON AGRICULTURE

Root-lesion nematodes (Pratylenchus) rip a hole in the sides of root cells and crawl inside. They move through the root, piercing, sucking, and leaving behind a trail of both cell-killing metabolites and eggs. Cell death results in brown lesions on the roots. Lesions begin on one side, but may encircle a root and thereby girdle it. The overall effect is a weak, shallow root system with a lot of dead areas.

Root-knot nematodes (Meloidogyne) enter roots as juveniles, select a feeding site of three to eight cells, and swell up in their chosen spot as they progress towards adulthood. They introduce hormone-like substances into the plant cells, causing the plant to swell in the area producing galls, or root knots. Males regain their slender profiles and leave the root at adulthood, but the fattened adult females remain inside.

They exude eggs into the soil, and after the juveniles hatch, the cycle begins again. Roots infested with root-knot nematodes usually have visible galls and may exhibit excessive branching. Parasitized plants may be weak and stunted. Root systems may be deformed, and underground organs such as potato tubers and carrot taproots may be damaged and unmarketable.

Pin nematodes (Paratylenchus) pierce root cells from the soil outside of the plant. Low numbers may appear inconsequential, but in high enough numbers, pin nematodes can damage crops. Ring nematodes (Criconemella) behave similarly. Little is known about ring nematode damage on many crop plants.

Stubby-root nematodes (Trichodorus and Paratrichodorus) are also external soil-dwelling parasites. Besides piercing root cells and sucking out the contents, they can vector viruses such as the Tobacco Rattle Virus, which causes corky ringspot disease in potato tubers. Other virus-vectoring plant-parasitic nematodes are Xiphinema, daggar nematode, which can vector Tomato Ringspot Virus, a disease of many fruit crops, and Longidorus ("mint nematode").

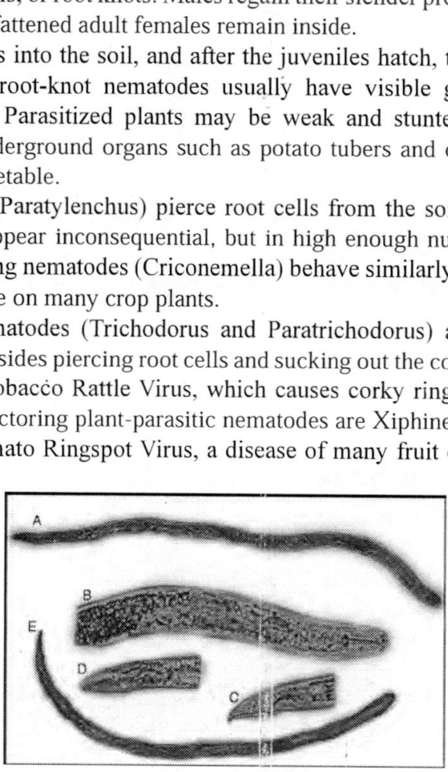

Fig. Vermiform (a) Adult Female Lesion Nematode, (b) anterior Region Showing Lobed Stylet, (c) Posterior of Female, (d) Posterior of Male Showing Spicule and Caudal Alae, (e) adult male lesion nematode

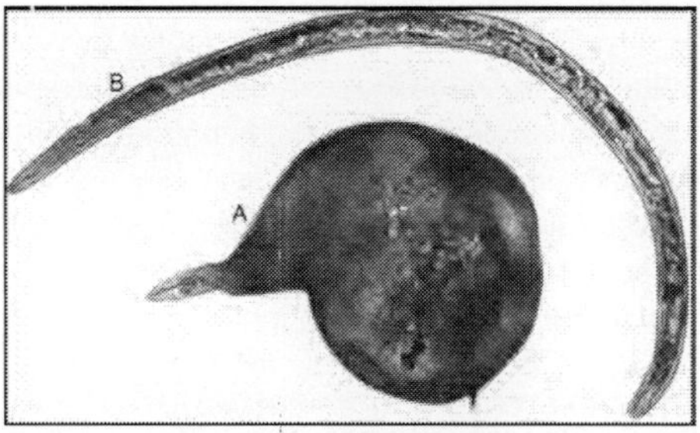

Fig. (A) Pyroform Female Root-knot Nematode, (B) Vermiform Male Root-knot Nematode

Nematode Organ Systems

Skeletal - exoskeleton (triple layered cuticle) Secretory/excretory -
Nervous - nerve ring and longitudinal muscles
Digestive - stoma, esophagus, intestine, Reproductive -
NO circulatory or respiratory systems - exchange of gasses through cuticle.

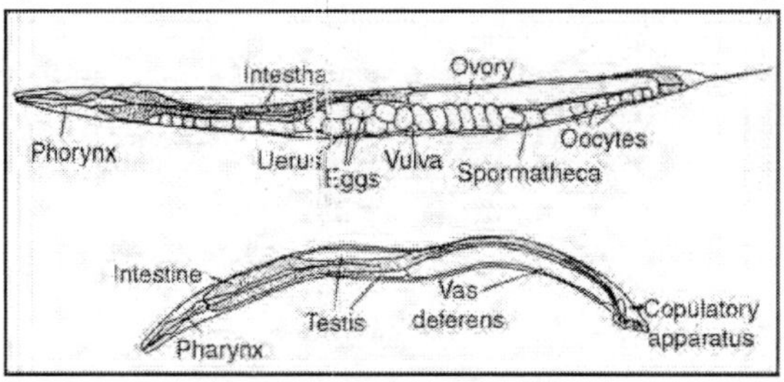

Fig. (Top) Hermaphrodite, (Bottom) Male

GENERAL PLANT PARASITIC NEMATODE LIFE CYCLE

5 Stages-4 Molts
Egg - J1 - M1 - J2 - (Hatch) - M2 - J3- M3 - J4 - M4 - Adult
In all plant parasitic nematodes the first molt occurs in the egg.
The emerging second-stage juvenile (J2) is the most common infective stage.
Length of life cycle of plant parasitic nematodes can be anywhere from 20-40 days (on average 25 days at 22oC).

NEMATODE MODES OF REPRODUCTION

Amphimixis - cross fertilization (males and females)
Parthenogenesis- nonsexual, males not required for reproduction(clonal)
Hermaphrodism - male and female organs in one individual(some change sexes)

Reproductive Potential (Egg production)

Anquina 25,000 eggs (2,000/wk)
Meloidogyne 500-1500 eggs C. elegans 250-350 eggs Ascaris 2 million eggs/day

NEMATODE GROUPS

1. Marine, Freshwater, Soil
2. Animal Parasites
   A. Vertebrate
   B. Invertebrate
3. Plant Parasites

Marine, Freshwater, Soil (Trophic Groups) Microbivores ("free-living") Fungivores Omnivores (feed on multiple sources) Predators (on nematodes, protozoa,etc.)

PLANT PARASITIC NEMATODES

PLANT PARASITES

197 Genera, 4300 species of phytoparasites Obligate parasites (cannot culture axenically, but can culture monoxenically)

EVOLUTIONARY ADAPTATIONS

Plant Parasitic Nematodes have protrusible, hollow stylet, and morphological and physiological changes to the esophagus. All plant parasitic nematodes have a protrusible, hollow stylet, BUT not all stylet bearing nematodes are plant parasites.

MOST DAMAGING PLANT PARASITIC NEMATODES

Root-knot Nematodes

- oot knot nematodes cause more damage worldwide than any other genus
- some species can parasitize more than 2,000 plant species
- form complex feeding sites in root called "giant cells"
- galls are formed on root
- reproduce by parthenogenesis
- eggs laid outside the female body in a gelatinous matrix sac that often protrudes to the surface of the root and can be seen on the root surface as a sign
- 500-1000 eggs laid per female

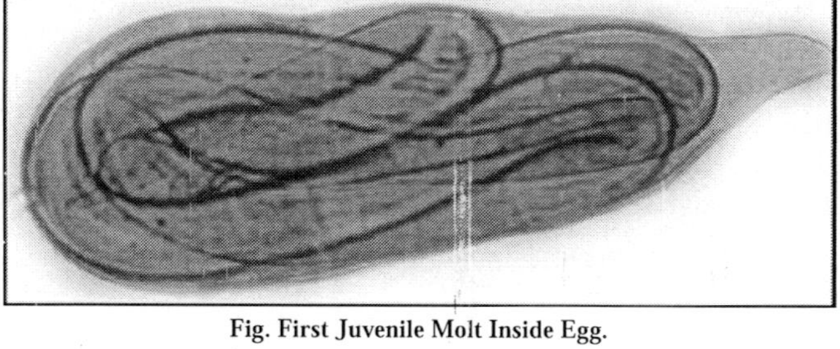

Fig. First Juvenile Molt Inside Egg.

At the establishment of a feeding site several dramatic physiological changes occur in both the host plant and the developing nematode including:

- Secretion of esophageal gland products possibly including hormones and enzymes required for feeding site formation and maintenance
- Degeneration of longitudinal muscles of J2 rendering nematode immobile(sedentary)
- Basal zone of the cuticle breaks down allowing for swelling during feeding
- Males third molt is a true metamorphosis seen in greater numbers under adverse environmental conditions by female sex differentiation(reversal)
- Undifferentiated vascular cells are targeted for feeding site initiation, most likely these are xylem parenchyma cells
- Feeding cells increase in size up to 100per cent(hypertrophy)
- Cell contents multiply including mitochondria and nuclei(40-100 per cell)
- Walls develop elaborate ingrowths and become thickened possibly to provide better absorptive abilities to meet the nutritional demands of the growing nematode • Giant cells degenerate upon nematode death, suggesting required stimulation by nematode for maintenance.
- J2 free in soil,

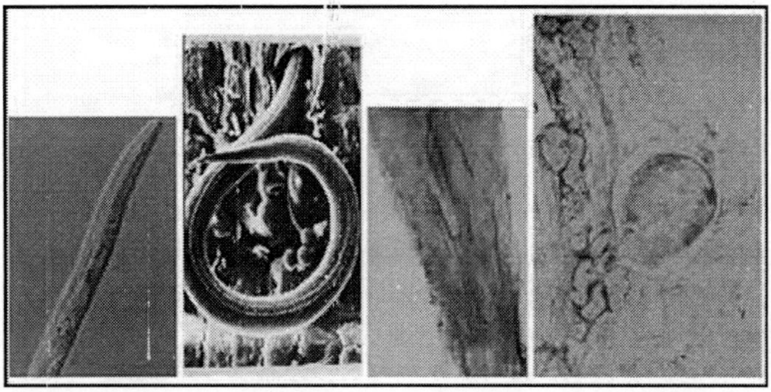

Fig. Infection Cycle of RKN.

- J2 penetrating root,
- J2s migrating to the vascular cylinder
- Adult female with 'giant cells' surrounding her head

Gall formation is a visible sign of root-knot nematode infection resulting from the hypertrophy and hyperplasia indicative of feeding site formation within the roots. (1975-1984) The International Meloidogyne Project was conducted to assess the worldwide impact of this genus on agriculture and was based here at NCSU.

Some of the results of this project include:

- The North Carolina Differential Host Range Test
- Frequencies of the most common species:

- M. incognita (Southern root knot nematode) 54per cent
- M. javanica (Javanese root knot nematode) 30per cent
- M. hapla (Northern root knot nematode) 7per cent
- M. arenaria (peanut root knot nematode) 7per cent

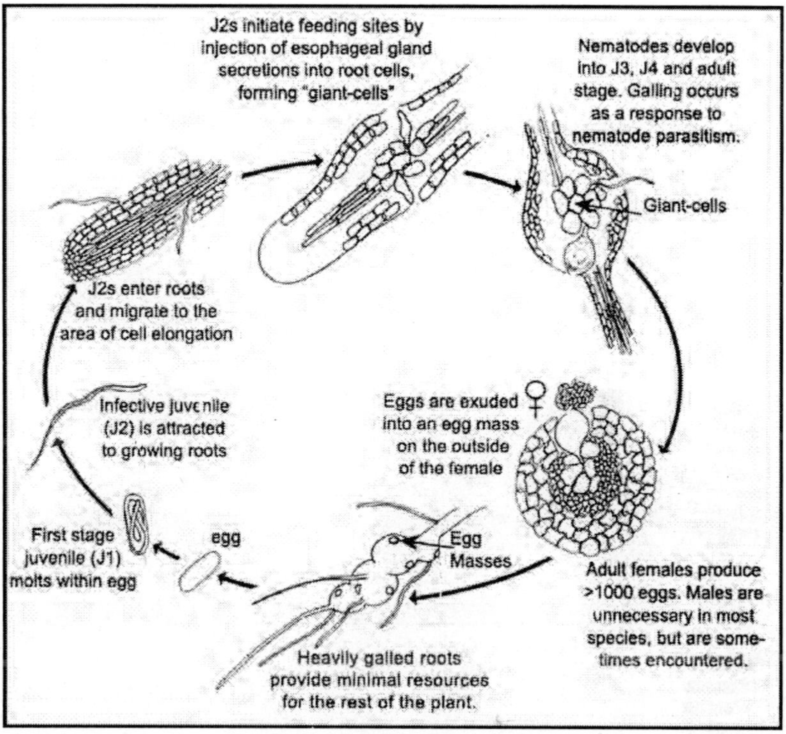

Fig. Root-knot Nematode Disease Cycle

The Cyst Nematode

Most Important Species

- H. glycines (soybean cyst - #1 pathogen of soybean)
- H. schachtii (sugar-beet cyst nematode; 218 host species)
- H. avenae (cereal cyst nematode)
- H. trifolii (clover cyst nematode)

General Characteristics

- Limited host-range.
- Reproduce by amphimixis.
- Form complex feeding sites called "syncytia".
- Cysts are lemon-shaped.

Life Cycle

- J2 cuts slit in egg shell with stylet(no enzymes involved)
- J2 migrates to root zone of elongation and penetrates to vascular cylinder
- Feeding site(syncytium) is initiated and J2 molts to sedentary J3
- Adult female produces 200-600 eggs that mature in the cyst
- Some species produce an egg mass for spontaneous hatching
- Eggs retained in cysts overwinter or survive up to 30 years(G. pallida)
- Hatching is often stimulated by hatching factors (root exhudates).

Hots Parasite Interactions

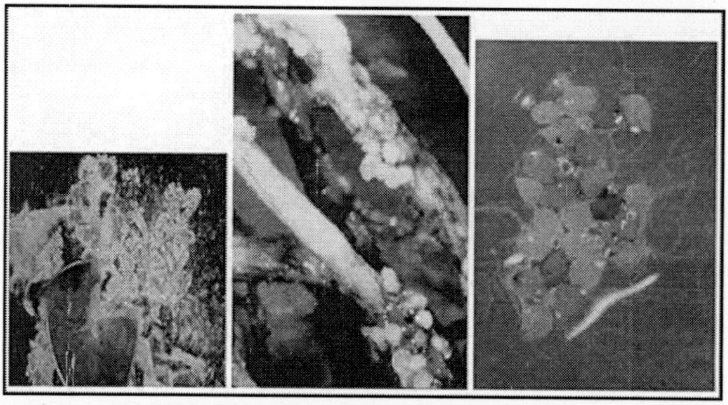

Fig. Left to Right: Broken Cyst with Eggs, Cysts on Roots, Juveniles Hatching from Cysts.

- Intracellular migration of J2 causes necrosis mechanically and with cellulases
- Syncitia are formed in pericyle and endodermal cells by dissolution of cell walls
- This syncytial initial is formed by the excretion of esophageal compounds
- Multinucleate cells result from the coalescence of neighbouring cells
- Nuclei enlarge and cytoplasm becomes granulated
- The syncitium shows increased metabolism in response to the nematode which is acting like a nutrient sink in the infected roots
- Feeding tubes form from nematode and plant secretions in the syncitium at the distal tip of the stylet, but are not enveloped by endoplasmic reticulum as in root-knot feeding tubes
- A feeding plug is also formed at the point of cell wall penetration
- No galling due to the lack of hyperplasia and hypertrophy
- Degradation of the syncitium results in considerably more necrosis than root-knot

Ditylenchus

Ditylenchus (The Stem and Bulb Nematode) migratory endoparasite mainly in shoots diverse feeding habits (can feed on fungi)

Important Species

- D. dipsaci: can infect over 500 plant species
- D. destructor: potato dry rot (mig. endoparasite of underground plant parts)
- D. myceliophagus: fungivore, pest for mushroom industry (France)

Life Cycle and Host-Parasite Interactions

- Reproduction by amphimixis (needs males)
- Females lay 200-500 eggs which usually hatch in the plant tissue
- All stages are infective, penetration at shoot apical meristem
- Migratory feeding leads to cavities in the parenchymous tissue
- Pectinase production is speculated in degradation of the middle lamella that results in cell separation and cavity formation during infection
- Swelling, galling, stunting, distortion, and necrosis result from serious infection
- Small numbers of nematodes are capable of causing serious damage
- No necrosis and little damage is seen in resistant varies with incompatible reactions
- 4th stage juvenile survival stage (up to 23 years)
- Produce "Nema Wool" an anhydrobiotic survival structure
- Cause galling and distortion to leaves, brown rings in bulbs
- Life cycle is 19-23 days at 15 degrees C

The Citrus Nematode

Most Important Species: T. Semipenetrans

- Only one species of economic importance in agriculture
- Limited host range (rutaceous plants, olive, grape, persimmon)
- Spread by infested nursery stock (which have no obvious symptoms)
- Causes "slow-decline" of citrus
- Semi-endoparasite (forms nurse cells in root)
- Young female is infective stage
- Long life cycle(6-8 weeks)
- J2 sexual dimorphism
- Develops to adult in 7 days without feeding

- 26per cent are non-infective males with esophoguses and stylets that degenerate
- 70-100 eggs laid in a gelatinous matrix
- 4000 nematodes/g root required to cause severe damage
- General symptoms are similar to nutrient deficiencies
- Symptoms not apparent until 5-10 years after population peak
- Feeds on 6-16 cortical(nurse) cells that show no hypertrophy or hyperplasia
- Syncytium has some wall ingrowths and the cytoplasm becomes granular with enlarged nuclei.

The Burrowing Nematode

- 31 species; 2 economically important (R. citrophilis, R. similis)
- Causes "spreading decline" of citrus/"toppling" disease of banana, and "pepper yellows" in black pepper
- Causes citrus yield reduction in Florida (up to 40-80per cent per grove)
- Quarantine procedures have prevented spread (no damage in California)
- Can feed at depths of 10-12 feet
- All stages are infective
- Reproduction by parthenogenesis with reports of facultative mphimixis
- Very prolific root colonization (2,500/inch of sugarcane roots)
- Can migrate from tree to tree through root grafts affected trees have little new growth, foliage loss, smaller fruit
- Intracellular migration leaves cavities in parenchyma tissue of cortex and stele
- Lesions appear on roots after 5-6 days turning brown and cracked in 15-20 days
- Some control provided by chemical and fallow barriers.

Dagger Nematode

Most Important Species: X. Index

- Major hosts: woody plants, tree fruits, soybean, corn, and some cereals.
- Sedentary ectoparasite feeds at or near root tips, may cause slight galling and necrosis.
- X. index is a problem on grapes and also vectors grapevine fanleaf (GVFL) virus.
- Large nematode, long stylet, feeds deeply.
- Very long life cycle: several months to 2 years.
- Causes "giant cells and galls" like RKN, and necrosis.
- Giant cells caused by karyokinesis without cytokinesis yeilding mutinucleate cells.

- Cessation of root growth caused by less than 12 hrs of feeding.
- Root swelling may continue after feeding has stopped.

Spiral Nematodes

Most Important Species: H. Dihystera and H. Multicinctus

Their common name comes from the shape of their body after death, a spiral.

- Most cosmopolitan nematode in the US (ubiquitous distribution)
- Achieve high numbers in soil, but rarely cause extensive damage
- Climatic range is temperate to tropical
- Main crop host is banana
- Average life cycle is 30-45 days
- Males are rare suggesting parthenogenesis
- Typically feed on only a few cell around head 4-6 cell layers deep into cortex *H. multicinctus* causes serious damage on bannana roots and corms in the tropics.

The Sting Nematode

Most Important Species: B. Longicaudatus

Belonolaimus can be one of the most devastating plant parasitic nematodes!!!

- Very few nematodes can cause severe damage(<10 per 100cc soil)
- Hosts include grasses, field crops, trees, seedlings
- Like sandy soils of southeast US
- Problem on turf
- Very long stylet used to penetrate deeply into cortex and vascular tissue
- Can cause seedling death, necrosis in apical meristem leading to stunted, coarse roots.

BACKGROUND INFORMATION

Plant-parasitic nematodes are nearly microscopic, worm-shaped animals virtually invisible to the naked eye when in the soil. They can cause significant plant damage ranging from negligible injury to total destruction of plant material. The severity of plant injury resulting from nematode activity depends on several factors such as the combination of plant and nematode species and prevailing environmental factors including rainfall, soil types, land contour, and culture practices.

Although a few nematode species feed on above ground plant parts, such as leaves, stems, flowers, and seeds, the majority of these parasites feed on underground parts of plants, including roots, bulbs, and tubers.

Because of this below ground, "hidden" feeding activity, nematode damage to plants cannot always be diagnosed readily.

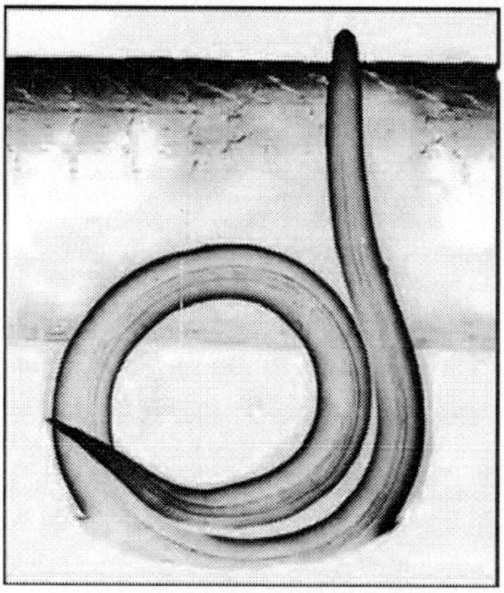

Fig. Scanning Electron Micrograph of a Second-stage root-knot Nematode Juvenile Propped up Against a Human Hair.

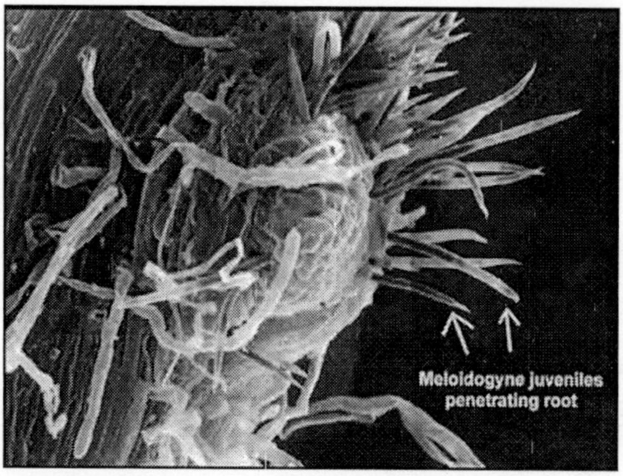

Fig. Meloidogyne Incognita Penetrating Cucumber Roots as Seen with Scanning Electron Microscopy.

Many plant-parasitic nematodes have a wide host range while a few are host specific.

- Nematodes also feed and reproduce on many different weeds.
- The survival of some nematodes for at least one-year in moist soil without a host plant is common.
- Many kinds of plant-feeding nematodes can survive freely in the soil through winter- even if the soil is frozen.
- A few plant-parasitic nematode species can survive in the soil in a dormant state within cysts for many years.

SYMPTOMS OF PLANT DAMAGE DUE TO NEMATODE ACTIVITY

Fig. Field Symptoms (Patchy Stunting and Chlorosis) of Soybean Cyst Nematode, Heterodera Glycines, on Soybean.

Initially symptoms are similar to lack of or improper fertilization, too little or too much water, improper soil pH, poor soil, or other environmental factors.

- Oval areas or irregular patches in the field with poor plant growth ranging from a few feet to hundreds of feet across.
- Poor, sickly growth, wilting, yellowing, stunting, and premature plant aging.

NEMATODE MANAGEMENT

- The individual situation determines the use of management tactics.
- The degree of plant damage varies with the kinds and numbers of nematodes present, the kind and variety of plant, and the cultural and environmental factors involved.
- Management tactics rarely destroy all plant-parasitic nematodes, but rather reduce the population(s) to levels below which economic damage occurs.
- When control measures are recommended, it is often necessary to repeat the measures periodically to maintain satisfactory plant growth.

Prevention: This strategy should be the first line of defence. Prevention is the practice of keeping a population of nematodes from infesting a site and specific tactics include:

- Education regarding nematode presence, biology, and management.
- Survey sampling of an intended planting site to determine the degree of existing nematode activity.
- Use of nematode free transplants.
- Careful Examination of transplant roots before planting. Plants can introduce nematodes to the site that attack present and future susceptible plantings.
- Weed control as a means of eliminating nematode access to alternate hosts.
- Sanitation practices such as cleaning of equipment between sites of operation.

Avoidance: practiced when nematode populations exist in a site, but their impact can be avoided or lessened through some cultural practice. Examples of avoidance tactics include:

- Sampling to determine nematode species and population levels.
- Choosing plant materials that are poor hosts.
- Practice crop rotations that include nonhost, resistant and susceptible crops when feasible.
- Employ cultural practices known to be optimal for plant maintenance.

Monitoring: includes tracking and proper identification of nematodes and includes:

- Soil and plant sampling for detection and trouble shooting.
- Routine sampling to maintain records of incidence and distribution.

These samples form the basis for crop rotations, cultivar selection, economic thresholds, suppressive actions, and other management decisions. Suppression: Nematode suppressive tactics may become necessary to avoid economic crop loss where prevention and avoidance tactics are not successful.

Suppressive tactics include:

- Addition of organic matter and green manure crops to encourage the release of microorganisms detrimental to nematode development or survival.

Chemical Management

- Chemical control involves the application of certain volatile (fumigants) and nonvolatile nematicides to the soil to reduce nematode populations. It is important here to note:
- There are some nematode problems for which there currently is no legal, effective nematicide.
- Chemical management reduces nematode populations for a limited period of time. Nematodes that escape treatment can resume feeding when the chemical is gone.
- Nematicides are severely restricted for landscape application.
- Because they are highly toxic, several nematicides that are registered for ornamental plants are restricted to commercial production.
- There is presently no effective nematicide that may be applied legally to ornamentals already planted in the landscape.

PARASITIC HIGHER PLANTS

Some higher plant forms live on the surface of or para-sitize other plants and often cause harmful reactions in their hosts. These plants can be placed in three groups: the epiphytes, the hemiparasites, and the true parasites.

EPIPHYTES

Epiphytes do little or no harm to their host plants, using them merely for physical support and protection. Examples are Spanish moss and epiphytic orchids, which in their native habitats commonly grow on tree limbs.

HEMIPARASITES

Hemiparasites, sometimes called water parasites, do injure their host plants, absorbing water and mineral nutrients from them. However, they possess chlorophyll and can manufacture their own carbohydrates by pho-tosynthesis.

Witchweed (Striga asiatica) is a hemiparasitic seed plant that severely damages sugarcane, corn, sorghum, many other grasses, and some broad- leaved plants. It attaches itself to the host's roots and utilizes most of the host's water and mineral nutrients, causing it to wilt, yellow, grow poorly, and die. The best control is to plant a crop, such as Sudan grass, that stimulates the witchweed seed to germinate, then plow under the entire field. Crops should be rotated and susceptible crops should not be planted.

Fig. Leafy Mistletoe on an Oak Tree.

Mistletoe (Phoradendromspp.) another member of the hemiparasitic group, attacks many broad-leaved trees such as ash, silver maple, honeylocust, hackberry, cottonwood, walnut, oak, birch, sycamore, and some conifers. The seeds germinate on limbs of susceptible hosts, forming an attach-ment disk on the bark. The sticky berries are disseminated throughout the tree and from tree to tree by birds and wind. The usual control, although not very effective, is to cut out the mistletoe branches deep into the tree proximal to the point of attachment. No good herbicidal control has been developed, although a dormant-season application of ethephon, an ethylene-releasing material, is a possible control on some plants. The best control is to plant only tree species resistant to mistletoe attacks.

TRUE PARASITES

True parasites lack chlorophyll and depend upon their hosts for all nourishment—carbohydrates, minerals, and water. Examples of this group are the dwarf mistletoe (Arceuthobium spp.) and dodder (Cuscuta spp.). Broomrape (Orobanche spp.)is a serious parasitic pest in Europe, Asia, and Africa and has caused extensive damage to tomatoes in California.

Dwarf mistletoe attacks many coniferous species in the western United States, reducing tree vigour and lowering lumber quality. The sticky seeds (not the fruits) are forcibly ejected and can travel up to about 19 meters (60 feet). This is the principal means of dissemination. Birds are known to carry the sticky seeds on their feathers, but wind plays a very minimal role in dissemination. The best control is removal of infected trees.

Dodder has many species, but about six cause the major damage, attacking such crops as alfalfa, lespedeza, clover, flax, sugar beets, potatoes, and some vegetable crops as well as ornamentals. Dodder seriously reduces yields and quality of crops. Strict regulations prohibit the sale of crop seed contaminated by dodder seed. Great effort should be taken to avoid planting seed that has dodder seed mixed with it. Patches of dodder in field crops or along fences or ditch banks should be eradicated by burning orby herbicides.

❐

9.

Plant Disease and its Classification

The study of plant diseases is known as plant pathology. Infectious diseases are caused by living organisms called pathogens. Non-infectious diseases caused by environmental stress and damage by weather and other environmental factors also will be covered. Indirectly, environmental factors that cause a plant to be stressed may result in the plant's gradual decline. Decline results in the plant being more susceptible to disease organisms.

Because of this, diagnosing plant diseases can be tricky. The real cause of a problem may be the stress factors, with the disease simply being a secondary factor.A plant disease is broadly defined as any condition in which a plant differs in some way from a normal plant in either structure or function. Plant diseases have had an important role in history. The holy fire in the Bible is believed to have been due to the ergot fungus in the heads of grain which made bread poisonous. The great famine in Ireland in the mid-1800s was caused by the destruction of the potato crop by late blight.

WHAT CAUSES PLANT DISEASE?

Plant diseases are generally divided into two groups based on their cause. Non-parasitic diseases are induced by some genetic or environmental factor such as nutrient deficiencies, extreme cold or heat, toxic chemicals (air pollutants, weed killers, or too much fertilizer), mechanical injury, or lack of water. These diseases cannot be transmitted to healthy plants and their control depends solely on correcting the condition causing the disease.

Parasitic diseases are caused by living organisms which derive their food by growing as parasites upon other plants. The most common causes of parasitic diseases are fungi, bacteria, viruses, and nematodes. A few seed- producing plants, such as the mistletoes, can also cause plant diseases.

Fungi are plants that lack the green colouring (chlorophyll) found in seed-producing plants and therefore cannot manufacture their own food. There are between

50,000 and 100,000 different species of fungi of many types and sizes, but not all are harmful.

Most are microscopic in size, but some, such as the mushrooms, are quite large. Fungi reproduce by spores. Bacteria are very small, one-celled plants that reproduce by simple fission. They divide into two equal halves, each of which becomes a fully developed bacterium. Plant diseases are frequently given their common names based on the plant part affected, by the way the plant is affected, or by the cause of the disease.

Viruses are so small they cannot be seen with an ordinary microscope. Many of the viruses that cause plant diseases are transmitted from one plant to another by insects, usually aphids or leafhoppers. Viruses are also very serious problems in plants that are propagated by bulbs, roots, and cuttings because the virus is easily carried along in the propagating material.

Nematodes are small eel-shaped worms that reproduce by eggs. The number of eggs produced by one female nematode and the number of generations in a season depends largely on soil temperature. Therefore, nematodes are usually more of a problem in warmer areas of the country. Most nematodes feed on the roots and lower stem of plants, but a few attack the leaves and flowers.

The life cycles of all disease organisms are greatly influenced by environmental conditions. Temperature and moisture are probably the most important factors that affect the severity of plant diseases. They not only influence the activities of the disease organism, but also affect how easily a plant becomes diseased and the way the disease develops.

How Are Plant Diseases Identified?

The extremely large number of plant diseases (about 50,000 in the United States) makes it impossible for any one person to be familiar with all of them. However, there are certain facts that help in the identification of plant diseases. The most important is to know the name of the plant that is affected. One can then check known plant diseases to see if a similar condition has ever been reported on that plant.

How the disease developed is also important. If it occurred in a very short time (overnight), it is probably not a parasitic disease, but is more likely due to some unfavourable environmental condition or chemical.

How is the condition distributed in the planting? Is it general over the area or only in one or two spots? Parasitic diseases usually do not affect a large percentage of the plants in the early stages, but start in one area and gradually spread to the other plants. And, parasitic diseases usually do not affect several different kinds of plants in one area at one time, even though some disease organisms can attack many different plants.

How Do We Recognize Plant Diseases?

The fact that diseased plants are in some way different from normal plants indicates that there are ways of recognizing plant diseases. Symptoms are outward expressions of plant diseases and include three general types: overdevelopment of tissue, such as galls; underdevelopment of tissue, such as yellowing or dwarfing; and death of tissue, such as blights and cankers.

How Are Plant Diseases Controlled?

The key to controlling parasitic diseases is to introduce an obstacle into the life cycle of the disease organism. This obstacle may be the use of disease- resistant plants on which the disease organism cannot grow. It may be to create environmental conditions that are unfavourable for disease development, or it may involve placing a protective chemical over the surface of the plant.

PLANT PATHOLOGY

DISEASE TRIANGLE

Three critical factors or conditions must exist for disease to occur: a susceptible Host plant, a pathogen, and the right mix of environmental conditions. The relationship of these factors is called the disease triangle.

If only a part of the triangle exists, disease will not occur. Understanding the disease triangle helps us understand why most plants are not affected by the many thousands of diseases that exist.

PATHOGEN

Most pathogens are host-specific to a particular plant species, genus or family. For instance, blackspot of rose will not attack marigolds or lettuce. Some diseases, such as the powdery mildews, produce similar symptoms on different plants. However, the fungi involved are usually host-specific. The rose powdery mildew fungus will not infect zinnias or turfgrass or vice-versa.

SUSCEPTIBLE HOST

A susceptible host has a genetic makeup that permits the development of a particular disease. The genetic defence against a disease is called disease resistance. This resistance can be physical characteristics of the plant (fuzzy or waxy leaf surfaces), chemical characteristics (enzymes that kill pathogens and lack of enzymes) and growth patterns (ability to block off diseased tissue or outgrow damage).

Plants also may be disease-tolerant. Even though infected with a disease, they can grow and produce a good crop or maintain an acceptable appearance.

The plant outgrows the disease and symptoms are not apparent or at a damaging level. It is important to remember that plants labeled as disease- resistant are resistant only to a particular disease. They are not resistant to all diseases. Resistance does not mean immunity. Under extreme circumstances, resistant plants may be infected by the disease to which they have resistance.

For disease to occur, the host plant must be at a stage of development that allows it to be susceptible to infection. For example, damping-off only affects seedlings. Botrytis is primarily a disease of buds, although it also can occur on flowers and leaves. Also, it is important that the pathogen be in a proper stage of its development to infect host plants.

ENVIRONMENTAL CONDITIONS

Certain environmental conditions must exist for disease pathogens to cause infection. The specific conditions vary for different pathogens. High moisture and specific temperature ranges, for example, are necessary for many fungal diseases. These conditions must continue for a critical period of time while the pathogen is in contact with the host for infection to occur. Moisture, temperature, wind, sunlight, nutrition and soil quality affect plant growth. If one of these factors is out of balance for the culture of a specific plant, that plant may have a greater tendency to become diseased.

For example, lilacs growing in shade are more likely to be infected with powdery mildew than those growing in full sunlight. Environmental conditions also affect the growth and spread of disease pathogens. Very dry or wet weather will have an accompanying set of diseases that thrive under these conditions.

Moisture

Moisture in the plant environment can include humidity, dew, rainfall or water from irrigation. Moisture is critical to the spread of most plant diseases. Familiar diseases, such as black spot, fireblight and apple scab require moisture to spread to and infect new host plants. Constantly wet foliage from overhead watering is a condition that promotes disease development. Seedlings grown indoors in soggy, unsterilized potting medium and pots are more prone to damping-off, a fungal disease.

Temperature

Each disease pathogen has a specific temperature range for growth and activity. There are warm-weather and cool-weather diseases. Many powdery mildew diseases are late summer, warmer temperature diseases. Temperature affects how rapidly pathogens multiply. Soil temperature can also be critical for disease infection.

Cool, wet soils promote fungal root diseases. Temperature extremes can cause stress in host plants, increasing susceptibility.

Wind and Sun

The combination of wind and sun affects how quickly plant surfaces dry. Faster drying generally reduces the opportunity for infection. Wind can spread pathogens from one area to another, even many miles. Wind and rain together can be a deadly combination. Windblown rain can spread spores from infected plant tissue, blowing these pathogens to new host plants.

Sunlight is very important to plant health. Plants that do not receive the right amount of sunlight to meet their cultural requirements become stressed. This may make them more susceptible to infection.

Soil and Fertility

Soil type can affect plant growth and also development of some pathogens. Light sandy soil low in organic matter favours growth of many types of nematodes. Damping-off disease increases in heavy, cold, water-logged soils. Soil pH affects pathogen development in some diseases. Clubroot of cabbage occurs in soils with a low pH, for example. High soil pH is a factor in the development of scab on potatoes.

Fertility affects a plant's growth rate and ability to defend against disease. Excessive nitrogen fertilization can increase susceptibility to pathogen attack. It causes formation of SUCCULENT tissue and delays maturity. This can contribute to certain patch diseases in lawns. Nitrogen deficiency results in limited growth and plant stress which may cause greater disease susceptibility.

DISEASE CYCLE

Often gardeners believe that their plants have become diseased overnight. This may be true in the case of damping-off. More often, however, much has occurred before symptoms are seen. There are five stages in disease development: inoculation, incubation, penetration, infection and symptoms.

The pathogen must be introduced (inoculated) to the host plant. Most pathogens cannot move on their own, but must be carried to the host plant. This is done by rain, wind, insects, birds and people.

Splashing rain carries spores of apple scab fungus from infected apple leaves to uninfected leaves. Wind blows fungal spores from plant to plant. The spotted cucumber beetle transmits bacterial wilt of cucumbers when feeding.

Working in the garden when plants are wet is a common way to spread disease. Disinfesting tools requires a 9-to-1 solution of water and bleach and takes a minimum of ten minutes. Smokers can transmit tobacco mosaic virus from a cigarette to tomato plants.

Seeds or cuttings from infected plants will also transmit disease. Certified seed guarantees that at the time of sale the seeds are free of all diseases. Seeds are often coated with a fungicide to prevent the transmission of surface fungal diseases.Disease-free stock guarantees that the plant is not infected with disease. This is particularly important with perennial plants, such as roses, raspberries and other small fruits.

Incubation

The second stage of disease development is incubation. The pathogen changes or grows into a form that can enter the new host plant. In many fungal diseases, the pathogen arrives on the plant as a spore which must germinate before it can grow into the plant.

Penetration

The third stage is penetration or the point at which the pathogen actually enters the host plant. Once the fungal spore germinates, it sends out thread- like tubes call hyphae. These penetrate the plant through wounds or natural pores. Wounding roots of bedding plants during transplanting provides entry for root-rotting fungi. The mouthparts of an insect also result in openings for penetration.

Infection

The fourth stage is infection. The pathogen grows within the plant and begins damaging the plant tissue.

Symptoms

As the pathogen consumes nutrients, the plant reacts by showing symptoms. Symptoms are evidence of the pathogens causing damage to the plant. Symptoms include mottling, dwarfing, distortion, discolouration, wilting, and shriveling of any plant part.

PATHOGEN SURVIVAL

Many pathogens can survive without a susceptible host even under the most unfavourable conditions. Many plant diseases survive from one growing season to the next on plant debris, seeds, alternate hosts or in soil.

Because of pathogen survival, it is important to remove and properly dispose of any infected plant materials. It is also important for the gardener to know about the diseases that affect each plant throughout the home landscape, as well as the conditions needed for proper culture.

PLANT PATHOGENS

Pathogens are grouped taxonomically into related groups. It is helpful to know the type of pathogen in order to prescribe a control. Plant disease names are often very colourful and descriptive, as in fireblight, black knot and clubroot. Many disease names indicate something about the appearance of the symptoms.

Fungi

Fungi (plural of fungus) are the largest group of plant pathogens. Fungi often develop into colonies, like bread mold. Some fungi develop into large structures, such as mushrooms. Fungi spread by spores carried by wind, water or animals. They enter host plants directly or through natural openings or wounds. They cause damage by producing substances that change or destroy plant tissues.

Common fungal diseases include powdery mildew, rust, leaf spot, blight, root and crown rots, damping-off, smut, anthracnose, and vascular wilts.

Powdery Mildews

These fungal diseases are very HOST-SPECIFIC. It is common to see powdery mildew on lilacs, roses, zinnias, melons, beans, cucumbers, turfgrass and many other plants. The symptom is a white or gray, powdery growth on leaves and stems.

Powdery mildew will not usually kill a plant; however, it may weaken plants, and it is unsightly. Providing adequate sunlight and air circulation to lower relative humidity and using fungicide sprays for high-profile plants help control powdery mildew.

Rust

Rusts are also host-specific. They produce masses of easily noticed, orange or dark red spore masses on leaf tissue. Rusts are commonly seen in our region on turfgrass, hollyhocks, snapdragons, hawthorns and the small fruits. Rusts usually develop during cool weather.

They are spread by wind and splashing water. Rusts require water to reproduce and infect host plants. Controls include watering early in the day and applying protective fungicidal sprays.

Leaf Spot and Blight

Rose growers are very familiar with black spot, a common leaf-spotting disease of roses. Scab on crabapple and pyracantha are leaf-spotting diseases that can cause defoliation. Botrytis flower blight occurs on annuals and some perennials during wet seasons. Anthracnose of shade trees is a blight frequently seen on sycamores, maples, ashes and oaks. Fungi-causing tree diseases are spread by wind and splashing water.

Grow resistant cultivars, maintain plant vigour, enhance leaf drying by proper pruning and water early in the day to help avoid leaf spotting and blights. Fungicides provide protection from infection when applied before infection by fungal spores.

Root and Crown Rots

These diseases, which include water mold diseases, attack a wide variety of plants. They cause damage to roots and crowns of plants.

Damage to some of the root system results in poor growth, yellowing or stunting of the plant. Root rot fungal pathogens are found in almost all soils. However, they do not survive as well in well-drained soils. These organisms can live in soil for years in a dormant state. Proper cultural practices, such as correct planting depth and improving drainage, are important controls.

Stem and Twig Cankers

Cankers may be compared to sores on a stem or twig. The tissue may be raised or sunken, discoloured or split open. Twig cankers on junipers and leaf and stem blight of pachysandra are common examples. Most of the canker diseases are not controllable with fungicides.

Maintaining plant vigour, protecting from injury, removing diseased parts and growing plants suitable for the site help prevent canker problems.

Vascular Wilts

Dutch elm disease is the classic wilt disease. The fungus invades the vascular system of the plant and clogs it, stopping the flow of water and nutrients. Verticillium wilt is common on trees, shrubs, vegetables and fruits. Discolouration of the vascular tissue sometimes can be seen when you cut diagonally through a branch. There are no effective chemical controls for wilts. Growing resistant species, maintaining plant vigour and sanitation provide the only controls.

Smuts and Molds

Smut fungi produce large amounts of black sooty spores which show up as unsightly dark brown or black blotches on leaves, stems and fruit.

Bacteria

Bacteria are microscopic, single-celled organisms that multiply rapidly. They produce chemicals that destroy plant tissue or cause abnormal growth. Bacteria enter through wounds, stomata or other natural openings. Bacteria can cause leaf spotting as on English ivy. Fireblight on crabapples, pears, mountain ash, hawthorn, cotoneaster and pyracantha is a bacterial disease. Bacteria also cause diseases, such as crown gall on stone fruits and euonymus and wilts in geraniums and cucurbits (cucumbers, squash and melons).

Disease-free stock and resistant cultivars plus sanitation help prevent bacterial diseases. When cutting out bacterial infections, sterilize pruners with 70 per cent alcohol between cuts. Streptomycin and fixed copper sprays may help slow the spread and development of bacterial diseases. However, chemical controls for bacterial diseases are less reliable than for many fungal diseases.

Mycoplasmas

Mycoplasmas are disease agents that were not accurately identified until the 1960's. Previously many mycoplasma-caused diseases were thought to be caused by viruses. A group of diseases caused by mycoplasmas is called YELLOWS diseases. Yellows disease is usually spread from one host plant to another by an insect vector. Leafhoppers are common vectors for mycoplasmas.

Viruses

Viruses are organisms so small that they can only be seen under an electron microscope, magnified 2,000 to 3,000 times. Viruses multiply only within living cells of host plants. Viruses are spread by insects, nematodes and humans. Symptoms include vein banding.

Mosaic, flecking or spotting on foliage and abnormal growth, similar to herbicide damage. Tobacco mosaic virus on tomatoes and leaf mosaic on dahlias are common viral diseases. Control is limited to removing and destroying infected plants.

Nematodes

Nematodes are microscopic, worm-like organisms. They most commonly feed on plant roots, but some nematodes invade leaf tissue. Nematodes suck out liquid nutrients and inject damaging materials into plants. They injure plant cells or change normal plant growth processes. Symptoms of nematodes include swelling of stems or roots, irregular branching, deformed leaves, lack of blossoming and galls on roots. Nematodes can facilitate the entry of viruses and fungi into plants. Control efforts are difficult and provide limited success. Chemical controls must be done by professional pesticide applicators. A specialized soil test for nematodes indicates the type of nematode.

DIAGNOSIS OF PLANT DISEASES

To correctly diagnose plant disease problems, follow a few basic steps. View the plant and its environment from various perspectives.

The most obvious place to look first is up close. Use a hand lens if necessary. Don't stop at the first or most obvious symptom; check for more. You want to find all of the symptoms. Look for symptoms on leaves, stems, roots, flowers and fruits. Cut open a branch or stem to look for vascular problems. Vascular problems show as discolouration of vascular tissue, leaf or stem wilting and sudden wilting of a section or a total plant.

General View

Stand back and look at the overall picture. Consider the total environment: weather, soil, stage of development for plant and pathogens, cultural practices and condition of other plants in the area. A plant growing in the wrong location may be stressed. Consider pesticide applications, recent construction or digging, and weather conditions.

Time

Determine when the symptoms became apparent. The onset of a problem may be due to a cultural practice, the seasonal appearance of a disease or insect, or a weather-related event. Remember that long-term stress is slow to appear, taking a year or more at times. Is the problem spreading? This may indicate it is a pathogen. Are plants of other species affected? Diseases are usually species-specific. Problems caused by environmental factors do not spread, although the symptoms may become more severe.

Knowledge and Experience

You must know what the plant should look like to be able to determine abnormalities. Check the references to see what problems are typical for a particular species. Gather all the information you can to help you make the diagnosis.

Remember that there is usually no single cause. There may be a primary cause; however, it may be associated with cultural or environmental conditions. Just as there is probably no single cause, there is usually no single symptom. Search for all of the symptoms. Orderly thinking and good questions are the key to accurate diagnoses. When in doubt about a diagnosis, turn to agents, state specialists or the Pest and Plant Disease Clinic for assistance or a second opinion.

Non-living or Abiotic Agents

Non-living or abiotic agents can indirectly result in plant problems. Additionally, several factors in the plant's environment can produce disease- like symptoms: weather extremes, high winds, high or low temperatures, nutrient deficiencies, physical damage and poor cultural conditions. Frost often damages buds and leaves in early spring. Hail can cause leaf spotting or holes. Drought and high winds result in wilting and in extreme cases, browning and curling. Air, water and soil pollution affect plant health and can produce disease-like symptoms. Soil imbalance, resulting from construction or other dumping, or misapplied garden chemicals can cause damage and disease- like symptoms.

Dumping of household, automotive and industrial chemicals can also produce plant damage. Plant disease can result from a combination of abiotic agents and biotic agents. Plants may be initially placed under stress by non- living agents. This creates a susceptibility in plants for attack by living agents. Drought may damage roots which then are more likely to be infected by fungal diseases.

Random or Uniform Patterns

Random distribution of symptoms on injured plants is usually caused by a biotic factor, such as infectious disease pathogens or an insect/animal. Uniform patterns are generally associated with abiotic or non-infectious agents like pesticides, fertilizers, environmental or site stress and mechanical damage.

Collecting Specimens

Plant specimens that are to be diagnosed should be taken from the area where symptoms are showing on living tissue. Dead plants are often invaded by secondary pathogens which may hide the original problem. Collect several representative samples showing various stages of disease development. A generous sampling will assist in diagnosis. If possible, collect the entire plant, including roots. Wrap the specimens in dry paper. Do not moisten them or seal them in plastic wrap or plastic bags. Never mix different specimens in a single bag. A fresh sample is required. Complete the diagnostic form as thoroughly as possible. This will result in better diagnosis.

DISEASE CONTROL

By the time disease symptoms appear, disease pathogens are inside the plant and usually beyond control. Therefore, it is important to prevent penetration of pathogens. Consider the following methods of disease control:

Avoid

Avoid certain diseases through choice of appropriate site and planting time, purchase of disease-free stock and cultural practices that do not favour disease infection. Insect control can be important for controlling spread of viruses, mycoplasmas, bacteria and fungi. Mulching can help control diseases by preventing the contact of foliage with soil.

Exclude

Exclude insect vectors with row covers. Don't bring diseased plants into the garden. Use Resistant Or Tolerant Plants Appropriate selection of species or cultivars will avoid or minimize problems. Remove all diseased plants, including alternate hosts. Immediate removal of diseased plants or plant parts reduces the chance of the disease spreading. Selective pruning and careful sanitizing of pruning equipment can prevent spread of disease. Rotate crops to avoid soil-borne diseases. Use best cultural practices. Follow recommended cultural practices.

TREAT

Treat plants with recommended chemicals to prevent infection. Some treatments will stop the spread of the pathogen.

Chemical Treatment of Plant Disease

To use fungicides and bactericides effectively, follow these 4 steps:

- Be sure that the diagnosis is correct.
- Choose a recommended control. Remember, a fungicide controls fungal diseases and bactericides only control bacteria. Read the product label and follow the instructions precisely. Make sure the plant you will be treating is listed on the label.
- Apply the control correctly following all label instructions. To be effective, fungicidal sprays must be applied properly. They work by creating a barrier that pathogens cannot penetrate. Good coverage is important. Give special attention to the underside of leaves, especially lower leaves. The barrier's effectiveness is affected by how well the spray spreads and sticks. Many fungicides have SPREADER-STICKERS added to the product. If you are not sure about coverage, observe the pattern spray deposits.

- Apply at the recommended time. In many cases, fungicides are a protectant and must be applied just before the pathogen is able to infect the plant. As the plant grows, additional sprays may be necessary to protect the new growth.

Fungicides are developed so that they degrade fairly rapidly. They are affected by rain, sun, and microbial activity. Excessive rainfall or growth will require shorter intervals between spraying. New shoots should be treated. Product label guidelines usually indicate a range of 7 to 14 days.

Know exactly what disease pathogen or combination of pathogens and abiotic agents are causing a plant problem. Fungicides and bactericides are effective against specific diseases under specific conditions. In the past, general fungicides were used. Many pathogens developed resistance to these general pesticides.

Use common sense in analyzing a situation. If an expensive or sentimental plant is experiencing problems, seek the help of professionals at the Pest and Plant Diagnostic Clinic.

PEST-RESISTANT CROPS

One of the mainstays of integrated pest management is the use of crop varieties that are resistant or tolerant to insect pests and diseases. A resistant variety may be less preferred by the insect pest, adversely affect its normal development and survival, or the plant may tolerate the damage without an economic loss in yield or quality.

Disease-resistant vegetables are widely used, whereas insect-resistant varieties are less common but non-etheless important. Examples include varieties of wheat which have tough stems that prevent development of the Hessian fly and cucurbits (squash, cucumbers, melons) that have lower concentrations of feeding stimulants (cucurbitacins) for cucumber beetles.

Commercial corn lines with increased concentration of a chemical called DIMBOA display resistance to European corn borer, a major pest throughout the world. In the case of cabbage, the only reliable method of controlling onion thrips is through the use of resistant varieties.

Advantages of this tactic include ease of use, compatibility with other integrated pest management tactics, low cost, and cumulative impact on the pest (each subsequent generation of the pest is further reduced) with minimal environmental impact.

The development of resistant or pest-tolerant plant varieties, however, may require considerable time and money, and resistance is not necessarily permanent. Just as insect populations have developed resistance to insecticides, populations of insects have developed that are now able to damage plant varieties that were previously resistant.

CULTURAL CONTROL

There are many agricultural practices that make the environment less favourable to insect pests. Examples include cultivation of alternate hosts (e.g., weeds), crop rotation, selection of planting sites, trap crops, and adjusting the timing of planting or harvest. Crop rotation, for example, is highly recommended for management of Colorado potato beetle. The beetles overwinter in or near potato fields and they require potato or related plants for food when they emerge in the spring. With cool temperatures and no suitable food, the beetles will only crawl and be unable to fly. Planting potatoes well away from the previous year's crop prevents access to needed food and the beetles will starve. The severity and incidence of many plant diseases can also be minimized by crop rotation, and selection of the planting site may affect the severity of insect infestations.

Trap crops are planted to attract and hold pest insects where they can be managed more efficiently and prevent or reduce their movement onto valuable crops. Early planted potatoes can act as a trap crop for Colorado potato beetles emerging in the spring. Since the early potatoes are the only food source available, the beetles will congregate on these plants where they can more easily be controlled.

Adjusting the timing of planting or harvest is another cultural control technique. The earlier planted processing tomatoes grown in the western United States are far less likely to be infested by the tomato fruitworm than those planted later in the season. It is also important to use pest-free transplants. Some vegetable crop transplants can be infested with insect pests, and growers using these transplants are put at a considerable disadvantage.

PHYSICAL AND MECHANICAL CONTROL

The use of physical barriers such as row covers or trenches prevents insects from reaching the crop. Row covers can help prevent early-season damage to cucurbits by cucumber beetles, and plastic-lined trenches are effective in trapping large numbers of dispersing Colorado potato beetles in the spring and fall.

Cold storage is also considered a physical control and, although it does not necessarily kill the insect pests, it at least stops their development and further feeding on the stored crop. Other methods include hand picking of pests, sticky boards or tapes for control of flying insects in greenhouses and various trapping techniques.

❒

10.

Plant Pathology: An Introduction

DEFINITION OF PLANT DISEASE

In nature for any variety or species of plant there is a set of environmental conditions which favours its optimum development. In nature, however, such an ideal situation seldom exists, and in consequence every variety of plant is subjected to the vicissitudes of the environment. When one or more factors in the environment become particularly unfavourable, the development of the plant is an in some way, and in comparison we growing in the usual range of environment, it a 2 Pears to have abnormal characteristics. Diseased plants are distinguished by changes in their structure or physiology of local price which are brought about by unfavourable environment or by one or another parasitic-agency.

There is no clearly defined line of distinction between normal or healthy plants I and abnormal or diseased plants. Plant disease, therefore, is, like many biological phenomena, difficult to circumscribe or define. We may, however, think of diseased plants as those which have become altered in their physiological and their morphological development to such a degree that signs of such effects are obvious. These external signs on the plant which are characteristic of a given disease are known as symptoms. Whether all deviations from the normal in structure and/or function constitute disease is largely a matter of definition and depends in any case upon what one chooses to dieting as normal for the particular variety or species in question.

Recognition of abnormalities in plants goes back to the early historical records. In general they were attributed to the acts of supernatural beings, to vicissitudes of weather, to insects, or to parasitic seed plants. Only since the middle of the nineteenth century have parasitic microorganisms been accepted as causal agencies in plant disease. In the last two decades of the nineteenth century viruses were recognized as infectious causal agencies, but only now has considerable evidence been brought forward to show that viruses are non- living but infectious proteins.

Since we now recognize that disease in plants may be brought about by unfavourable environment, or by infectious agencies which mayor may not be living, it is well that we define at the outset certain terms which will be used repeatedly in this book.

COMMON DEFINITIONS

First of all it is well for the student to have a clear concept as to what constitutes the cause of disease. This is particularly true because the term is used rather loosely in much of the literature on plant pathology. It is relatively simple to conceive of one or more, or a combination of several, environmental factors becoming the cause of a disease when they alter the physiological processes and morphological development of the plant concerned. When a parasitic microorganism enters into the complex, as is the case in a large percentage of plant disease, there is a common tendency to refer to the microorganism as the-cause of the disease.

Since we now know that variation in the environment is sometimes necessary to predispose the plant, i.e., cause it to become susceptible to the microorganism, one can readily see that the microorganism is not the sole causal factor. Moreover We also recognize that the microorganism, also entity, is subject to the influence of environment, and it often, if not always, requires a certain range of favourable environment to produce infective organs and for such organs to infect the plant. After a microorganism has successfully infected a susceptible plant, the subsequent interaction of plant and microorganism which results in a disease of the former is also environmental influences which determine whether disease develops at all and if so, at what rate and to what degree. It is evident that in the case of a disease in which a bacterium or fungus is concerned there are various other factors besides the latter which enter into the causal complex.

Strictly speaking, therefore, it is not correct to refer to a given microorganism as the cause of a given disease, since this implies that it is the sole cause. However, such usage is common in current literature on the subject. It has been suggested that the organism be referred to as the incitant, implying that it incites the disease under the influence of other factors. The term causal organism is also proper since it implies that the organism is a part of the causal complex.

Most causal organisms are parasites. In current usage a parasite, as applied to an organism, is one which subsists in whole or in part upon living tissue. Conversely, a saprophyte is an organism which subsists upon dead organic matter and inorganic materials. A pathogen is an agency which incites disease. While the term might be applied to any or all of the causal factors described above, it is practically always used to denote the living entity in the causal complex. The terms pathogen and parasite are not synonymous, however. In some cases organisms are parasitic without becoming causal factors in disease.

In other cases the by-products of a Retract saprophyte are an important part of the causal complex, and thus a sarrophyte may Indirectly become a pathogen. Pathogenicity is the property of a microorganism whereby it may become a part of the causal complex. Pathogenesis is the process or chain of events whereby disease development takes place.

A not uncommon habit among plant pathologists is to refer to the disease concerned by the scientific name of the causal organism. This is, of course, strictly incorrect. The student of palp to pathology should at the beginning develop a habit of referring to the disease and to the causal organism as distinct entities.

Diseases of econorllic plants have usually acquired one or more common names. As a rule such names are descriptive of some important phase of the disease, e.g., late blight of potato wilt of tomato; black stem rust of wheat; club root of cabbage. Sometimes the generic name of the causal organism is used as a descriptive adjective in the common name, either because a sufficiently descriptive common word is not available or because it is necessary to distinguish between similar appearing diseases with different causal organisms. Thus wilt of tomato may be the result of the action of distinct causal organisms; hence, Fusarium wilt of tomato, Verticillium wilt of tomato, bacterial wilt of tomato.

The Rhizoctonia disease of potato is another example of a case where the genus of the parasitic phase of the causal organism is used in the common name of the disease. This organism has a wide range of pathogenetic effects upon the potato plant. Under certain conditions basal stem cankers are formed, the result of which may be upward rolling of leaves, abnormal purpling of the foliage, deposition of starch in axillary buds of the stem to form aerial tubers.

Quite apart from the disease on aerial parts of the plant, there are formed superficial black resting bodies on the surface of the potato tuber. Because the last symptom is a very obvious one, the disease is often referred to as black scurf. However, since- the last term describes only one phase of a wide range of distinct disease reactions, the first termentioned, Rhizoctonia disease, is also used. The causal organism, if fully described, has been given a specific name within a given genus according to the accepted system of plant nomenclature. Since the generic name and specific name are Latin words always printed in italics, they should always be underlined in any handwritten or typed manuscript. This name of the organism is referred to as the Latin binomial of the organism.

The species as originally described may be later transferred to another genus by the same or another investigator. In such a case the species name is retained unless it has been declared invalid for some legitimate reason. The name, or an abbreviation thereof, of the investigator who first described the species follows the binomial. If the species is referred by another investigator to another genus, unless the transfer is from a form genus to that of a perfect form, the name or abbreviation of the one making the original description is retained in parentheses immediately after the binomial, while the name or abbreviation of the one responsible for the current binomial follows.

Since there is not always agreement as to the proper genus to which a species belongs or as to whom the first authentic description of a species is properly attributed, there may be found more than one binomial for a given organism in current literature. On this account it is customary in a paper dealing with a given disease to give the complete binomial with the proper authority when the organism is first mentioned. This indicates clearly to the reader which authority the writer chooses to follow. Subsequently in the paper the authority may be omitted, and some writers prefer to make subsequent reference to the causal organism by using the common name of the disease as a descriptive phrase modifying "organism," "fungus," "bacterium," etc., e.g., the crown-gall organism, the potato late-blight organism, the potato-late-blight fungus, the fire-blight bacterium.

In certain cases usage has resulted in exceptions to the general rule, in that the same word is applied to the disease and to the causal organism. Most common examples are found in the rusts. The large group of diseases known as rusts acquired this name long before organisms were recognized as causal agencies, and the term came into use because it described a prominent characteristic of the diseases as a group. Like wise the organisms were known, and many were described and named, long before they were recognized as pathogenic agencies. Because of their characteristic colour and the consistency of the spore mass, this group of organisms was also referred to as rusts. Thus the same word came into use for disease and for pathogen long before their relations were fully understood. The same chronological history is true for the smuts. It is the more important, therefore, to connote whether the term is being used to apply to the disease or to the causal organism.

In the case of diseases in which parasitic organisms are a part of the causal complex the diseased plant is commonly referred to as the host. While usage has made this the most widely used term, objection has been raised that it does not imply the actual relation which ordinarily exists between the plant and the pathogen. The term suspect is a newer word coined to apply to the plant which is subject to a given disease brought about by a given causal complex. The above terminology does not always strictly apply in the case of virus diseases. While the inanimate nature of viruses is not accepted in all quarters and practically nothing is known about their mode of increase or reproduction, they have above all in common with pathogenic microorganisms the property of being transmitted from plant to plant to become a part of the causal complex. While several schemes of nomenclature for plant viruses have been proposed, none has been fully accepted as yet. Therefore in current literature the student will find a great diversity among authors in method of reference to a given virus. They are not universal referred to as pathogens, and their disease-producing property is not commonly referred to as pathogenicity, largely because of the doubt that they are living entities.

One system known as the Holmes system sets up a scheme of genera and species whereby each described virus is given a binomial. While this system has many admirable features, it is non-conformist in that it uses system set up for living organisms to classify what may well be non-living organic compounds, which cannot be defined on the bases used in the binomial system for living organisms. An earlier suggestion by Johnson is to give each virus described from a given suspect or host consecutive numbers, e.g., potato virus

1, potato virus 2, cucumber virus 1, etc. A modification of the Johnson system, proposed by Smith, substitutes the generic name of the host for the common English name, e.g., Solanum virus 1, Solanum virus 2, Cucumis virus 1, etc. Some writers follow one, some another, of these systems, while still others avoid all I three systems and refer to the virus concerned by means of the disease with which it is associated, e.g., common-tobacco-mosaic virus, potato ring-spot virus, tobacco-ring-spot virus, spotted-wilt virus, potato-witches broom virus, etc.

When a pathogen is a part of the causal complex, the organism is first present in the environment in some overwintering form. The overwintering stage may be in or on seed of the host, in perennial wild hosts, in debris of infected plants of the previous season, as dormant spores, sclerotia, or other organs in the soil, in the form of an inhabitant of the soil flora, and in various other forms. Whatever this source, it becomes the source of primary inoculum. It may serve as the infective agent and thus become the primary inoculum, or it may multiply as a saprophyte to produce spores which become the primary inoculum. When the host becomes diseased, the organism commonly produces another crop of spores or infective bodies, which are known collectively as the secondary inoculum and serve to cause secondary infection. This last cycle may then be repeated several times during the current season.

The first step in the relation of the inoculum to the host consists of penetration. This may take place in one of a number of ways which are characteristic of the organism and are influenced sometimes by the environment. The various methods- of penetration will be discussed in later chapters. It is important to recognize here that there is a distinction between penetration and infection Penetration refers to initial invasion of the host by an organism. We know that many organisms penetrate cells of plants which are not susceptible and which do not respond with any sign of disease; moreover they die without the stage of penetration. Thus we may have many cases of penetration without infection. Infection implies the establishment of the pathogen within the host following penetration. It does not imply the production of disease symptoms. There may be a long period between infection and appearance of disease, but ordinarily one follows rather closely upon the other. The interval is known as the incubation period. Just as penetration and infection are to be regarded as distinct, so infection and disease development are distinct. Disease development refers to the events which occur between the time of infection and the complete expression of disease. In general, all disease development is a protracted process which involves production of secondary inoculum and progression of external evidences of disturbance of the morphological and physiological processes of the host plant; these external evidences are referred to as symptoms. The chain of events in disease development is referred to as the disease cycle. While, when a pathogen is involved, the disease cycle is intimately associated with the organism, it is distinct from the life cycle of the organism. In the case of many causal organisms only part of the life cycle is passed in association with the host, the remainder being in association with dead host tissue, independently as a saprophyte, in association with an alternate host, or in association with an insect vector.

An infectious organism is one which may be transmitted from a diseased host plant to a healthy plant and which is capable of inciting the disease under favourable environment. An infectious disease is one in which a transmissible causal organism is a part of the causal complex Virus diseases are also usually referred to as infectious since the incitant agency can be transmitted from a diseased to a healthy plant and.initiate the disease in the latter. In the light of present knowledge it is not proper to group them with infectious organisms, since there is no evidence that they reproduce themselves, whereas it is entirely possible if not the infective entity acts as a catalyst which stimulates the formation of virus protein instead of normal host protein during the period of development.

Various levels of parasitism among microorganisms have long beer recognized. In many cases two living organisms may live in association with one another, and in a sense each may be parasitic upon the other However, they are not only non-pathogenic to one another but also are each essential to the other's development or at least each is beneficial to the other's development. This phenomenon is known as symbiosis and the participating organisms are known as symbionts. Lichens are classic examples of symbiosis a fungus and an alga living in close association with one annotate. The alga thoughts photosynthetic property converts kinetic energy of the sun's rays into potential energy of carbohydrates useful to its own development and essential to the life of the fungus. The latter through its property of producing hydrolytic enzymes breaks down materials in the substrate, making energy available and providing mineral and organic nutrients essential to itself and the alga. Another case of symbiosis is that of the nodule-forming bacteria in relation to their respective leguminous hosts. Here the bacterium penetrates and infects the root hair; the host responds by abnormal proliferation of cells to cause a gall, or to this point the process might be considered pathogenic. However, the bacteria are capable of fixing the free nitrogen of the air and converting it into the essential nitrogenous compounds in their cells. Presumably, as the bacterial cells disintegrate, the nitrogenous compounds become available to the legume plant. Since higher plants are generally incapable of utilizing free nitrogen, an element essential to protein formation, this symbiotic phenomenon is a great economic source of available nitrogen.

That symbiosis is a phase of parasitism and is very close to pathogenesis is shown by further study of the legume-bacteria phenomenon. All strains of the bacterium adapted to a given legume host are not equally efficient in nitrogen assimilation. In fact, some strains show little or no evidence of such a property. When such a strain is allowed to infect its legume host, nodules develop as usual, but in the absence of nitrogen assimilation the usual benefits to the host are lacking, and the latter shows certain disease symptoms when compared with uninucleate plants. Thus the legume bacterium is in this case not symbiont but a pathogen, the parasitism involved results in pythonesses rather than symbiosis.

Contiguous to parasitism but distinct from it is the phenomenon referred to here in as eplphytlsm. An epiphyte is a plant growing upon another plant, and, as the term is usually applied, it refers to cases where no peridotic or truly symbiotic relation is involved. Particularly in tropical religions, many higher plants grow on shrubs or trees, not in association it the soil. They secure essential mineral nutrients from the dead bark issue of the substrate plant but do not penetrate the living tissue of the apparently, have no beneficial or detrimental effect upon the Oscillator.

While some parasitic microorganisms are largely epiphytic, they are not considered as members of this group. Opposed to epiphytes are endophytes, which are contained entirely within the substrate plant.

Having delineated symbiosis and epiphytism in relation to parasitism, we may now consider various grades of the last. After the relation of microorganisms to plant disease became fully recognized shortly after the middle of the nineteenth century, it soon became-apparent that pathogenic organisms vary widely in the manner in which they bring about pathogenesis under the influence of favourable environment. Parasitism is a very complex phenomenon which is basic to the understanding of plant disease.

Its various phases will be referred to repeatedly in subsequent chapters. Here it is essential only to explain the important terms used throughout the book. Some pathogens, so far as is shown, cannot complete their life cycles except in a parasitic relation to the respectively. They may persist as dormant spores away from the host and such organs may germinate and produce secondary spores away from the host.

Thus part of the life cycle may proceed apart from the parasitic relation. Infection must occur, however and parasitism obtain for the completion of the life cycle. Such organisms are known as obligate parasites, and they are Bound in various groups of the fungi one are known among the bacteria.

Viruses have properties in common with obligately parasitic organisms in that they increase in amount only in association with living suscepts. Common examples of obligate parasites are the rust fungi the powdery-mildew fungi, the downy-mildew fungi, the crucifer-clubroot organism, the potato- black-wart fungus. Obligate saprophytes are those microorganisms which have no relation with living cells and secure their nutrients from dead organic tissue or from available inorganic materials. A large share of the pathogens, however, have parasitic and saprophytic phases of their life cycles. The fact that an organism is in part a saprophyte does not in any sense detract, necessarily, from its potentiality as a pathogen. Many organisms can live for indefinite periods as saprophytes in pure culture or otherwise and still remain highly virulent pathogens.

There are, however, many organisms which are in the main saprophytes but have weakly pathogenic properties which an ordinarily expressed only when particularly unfavourable conditions have predisposed the host in such a way that it is unusually susceptible and when such environment is particularly favourable to the expression of the weakly pathogenic properties of the organism. Such organisms are called facultative parasites, which term implies that they are ordinarily saprophytic but under certain conditions may become parasitic. Many damping-off organisms are typical facultative parasites.

At the other end of the wide range of parasites that are not obligate parasites in the strictest sense are the facultative saprophytes. These organisms ordinarily have life cycles similar to the obligate parasites, in that they pass most of the cycle in association with the host and do not subsist as saprophytes. They may, however, subsist as saprophytes for part of their life cycle but do not complete their life cycle away from the host. Many of the smut fungi answer the requirement of this definition of a facultative saprophyte. It should be borne in mind, however that there are no well-defined limits between facultative saprophytes and facultative parasites. It should be emphasized that they represent the extremes in the range of microorganisms which may have parasitic and saprophytic phases in their life cycles and particularly that, as defined, the facultative saprophyte has a much higher potentiality as a pathogen than does the facultative parasite.

Predisposition may be defined as the effect of one or more environmental factors which makes.a plant vulnerable to attack by a pathogen. It is a process which antedates penetration and infection. Susceptibility refers to a condition of a plant in which it is normally subject to attack by a given pathogen. The susceptibility of a given individual, variety, or species may be increased or decreased by environmental factors. It may also be higher or lower, as between individuals, varieties, or species because of differences in the inherited characteristics which affect susceptibility. Resistance may be regarded as the counterpart of susceptibility and also may be influenced in its expression by environmental factors.

Basic Classifications of Plant Diseases

Plant diseases may be grouped in various ways. The method of classification depends to a considerable extent 'upon the purpose of the treatment. In the chapter on History of. Plant Pathology it will be seen that before the acceptance of microorganisms as "causative agencies in plant disease many writers speculated more or less on possible causal factors, but they based their schemes of classification upon effect rather than upon cause.

After 1850 books upon plant diseases grouped them according to causal factors rather than according to effects. Classification on the basis of cause has many advantages in generally text, one of the chief of which is that the most satisfactory approach to a basic understanding of disease is through an unraveling of the various causal factors and their interrelations, as well as their interaction, with the host. For this reason such a scheme is used in this book.

In a manual of diseases of a certain group of plants, the consideration of classification of diseases according to the host plant has decided advantages, and such a scheme is often used, e.g., cereal diseases, truckeprop diseases, corn diseases, flax diseases, etc. In other treatises diseases advantageously treated according to symptoms or effects, e.g., root powdery mildews, wilts, leaf blights, overgrowth, or hyperplastic diseases, hypoplastic diseases, necrotic diseases, etc.

Parasitic and virus diseases are often classified, in occurrence, into three groups: endemic, epidemic, or epiphytotic, and sporadic.

ENDEMIC DISEASES

When a disease is more or less constantly present from year to year in a moderate to severe form, it is classed. This is usually based on the fact that the pathogen or virus is well established and long persistent through its ability to survive from one crop, season to the next in soil, on crop 'refuse, or in wild hosts.

It is also implied that the environmental conditions are usually favourable for inoculum to develop infection to take place, arid disease development to ensue. Diseases due to pathogens which persist for long periods in the soil, such as those of cabbage yellows and onion smut, are commonly endemic. When the infectious agent readily survive and in prevarlent wild hosts, as is the case with cucumber nomic in milkweed or pokeweed plants, the disease is also usually endemic.

EPIDEMIC, OR EPIPHYTOTIC, DISEASES

An epidemic disease is one which commonly occurs widely but periodically. The term "epidemic" is derived from a Greek word meaning "among the people" and in the strictest sense applies to diseases of man. The term "epiphytotic" has been coined to' carry the same connotation with reference to plants. Epidemic diseases are often very responsive to variations in environment. The pathogen may be more or less constantly present, as in endemic diseases, but the environment favourable to development occurs only periodically.

It is thus evident that the environment in such a case may well be the major determining causal factor. Epidemic diseases "may also occur where the environment is usually favourable but where the pathogen is irregular in its occurrence. This may be because of the inability of the organism to survive always from one crop to the next. It 4 may be dependent upon seed or other propagative parts of the host for introduction. Insect vectors may be necessary for introduction or spread Epidemics of black rot of crucifers are commonly dependent upon introduction of the pathogen with the seed. Late blight of potato is common epidemic in nature because of its sensitivity to fluctuation in temperature and humidity. Curly top, a virus disease of sugar beet in the Rock Mountain states, is dependent upon the introduction of the virus by hoppers which overwinter in more or less distant areas. If the winte reduces greatly the overwintering insect population, infection and spread are minor; if the conditions are the reverse, early and continued infection results in severe disease development.

SPORADIC DISEASES

Sporadic diseases in reality belong to the epidemil group. The term is applied to those which occur at very irregular interval: and locations and in relatively few instances.

It should be borne in mind that a given disease may be endemic in region and epidemic in another. Late blight of potato is endemic is certain parts of northern Europe because the environment is practical always favourable at some time during the season. It is epidemic in the upper Mississippi valley and on the winter crop in the Gulf states, because some seasons are entirely unfavourable for its initiation and development.

Diseases Classified According to Major Causal Agents

There are two methods by which major disease classes may be set up according to cause. By one method, diseases are divided into two classes infectious and non-infectious. In this case all diseases incited by microorganisms, parasitic phanerogams, nematodes, and viruses are regarded as infectious. However, it may be questioned whether certain of the insect- incited diseases are infectious, although they are parasitic. By the second method, diseases are divided into non-parasitic and parasitic classes. In this classification some authorities would place virus diseases in one class, and some would place them in the other.

I. Non-parasitic diseases: Diseases incited primarily by,
   1. Low temperatures.
   2. High temperatures.
   3. Unfavourable oxygen relations.
   4. Unfavourable soil-moisture relations.
   5. Injurious atmospheric impurities.
   6. Lighting injury.
   7. Mineral excesses.
   8. Mineral deficiencies.

II. Parasitic diseases: Diseases incited by,
   1. Bacteria.
   2. Plasmodiophorales.
   3. Phycomycetes.
   4. Fungi Imperfection.
   5. Ascomycetes.
   6. Basidiomycetes.
   7. Phanerogams.
   8. Animal parasites.
   a. Nematodes.
   b. Insects.

III. Virus diseases.

11.

Fundamentals of Organic Farming

PRODUCING AND EXPORTING ORGANIC PRODUCTS TO MAJOR MARKETS

DEFINITION OF ORGANIC FARMING

Products labelled as "organic" are those certified as having been produced through clearly defined organic production methods. In other words, "organic" is a claim on the production process rather than a claim on the product itself.

Organic agriculture is best known as a farming method where no synthetic fertilizers and pesticides are used. However, this description does not mention the essence of this form of agriculture, which is the holistic management of the farming system. According to the definition of the Codex Alimentarius, "organic agriculture is a holistic production management system which promotes and enhances agro-ecosystem health, including biodiversity, biological cycles and soil biological activity. It emphasizes the use of management practices in preference to the use of off-farm inputs, taking into account that regional conditions require locally adapted systems. This is accomplished by using, where possible, agronomic, biological and mechanical methods, as opposed to using synthetic materials, to fulfil any specific function within the system." Several national governments and a multitude of private certification and farmer organizations have defined organic agriculture. In the past, differences in these definitions were significant, but the demand for consistency by the trade has led to greater uniformity.

BASIC CRITERIA FOR THE PRODUCTION, STORAGE AND TRANSPORT OF ORGANIC FRUIT AND VEGETABLE

One of the essential elements distinguishing organic farming from other forms of sustainable agriculture is the existence ofproduction standards and certification procedures. There are no universal standards for production and handling of organic fruit and vegetables. Initially, organic standards were developed by private associations, entitling members to use the respective associations' organic brands and labels when marketing their products.

The International Federation of Organic Agriculture Movements (IFOAM), a non-governmental organization promoting organic agriculture internationally, has established guidelines that have been widely adopted for organic production and processing.

These guidelines are commonly considered as "minimum standards", leaving room for more detailed requirements, depending on regional or local situations.

As organic agriculture has become more widespread, many developed countries have defined their own organic standards. Since the early 1990s, EC countries have endorsed a common organic standard which is spelled out in Regulation EEC 2092/91. More recently, Canada, the United States and Japan have adopted organic standards and regulations. The Committee on Food Labelling of the FAO/WHO Codex Alimentarius Commission adopted "Guidelines for the Production, Processing, Labelling and Marketing of Organically Produced Foods" in 1999. Organic standards are usually similar as they derive from IFOAM's guidelines for organic production. In general, the use of methods that contribute to maintaining or enhancing soil fertility is mandatory.

Another common feature is that generally natural inputs are approved and synthetic inputs are prohibited. Nevertheless, there are exceptions in both cases. Certain natural inputs which various certification programmes determine to be harmful to human health or the environment are prohibited. In addition, certain synthetic inputs are allowed.

For example, EEC Regulation 2092/91 allows, when required, the use of specific fertilizers and plant preservatives. All certification programmes maintain lists of specific approved synthetic inputs and prohibited natural inputs. Many certification programmes require additional environmental protection measures beyond the above prerequisites.

For example, specific measures are generally applied in the areas of soil and water conservation, pollution control or the use of biological control agents. While many farmers in the developing world do not use synthetic inputs, this alone is not sufficient to classify their products as organic.

Farmers who produce organic crops for export and at the same time grow basic foodstuffs on other areas of the farm using conventional methods with fertilizers and pesticides, inadmissible to the organic system, are at high risk of violating the standards, unless effective measures are taken to prevent prohibited substances from passing to the organic plots.

CERTIFICATION: A NECESSARY STEP FOR EXPORTING FRUIT AND VEGETABLES AS ORGANIC

Basic Definition

Producers and exporters of organic fruit and vegetables seeking to sell their products in developed countries which have adopted organic standards and regulations will have to meet the rules established by the importing country concerned. In the United States, Japan and the EC countries, the largest markets for organic products, the standards applying to domestic organic production also apply to imported organic products. There are detailed regulations governing the production, importing, marketing and labelling of organic products.

Producers and exporters wishing to export fruit and vegetables with the organic label will have to obtain organic certification. Organic certification is a procedure for verifying that the production process conforms to certain standards. In other words, certification is primarily an acknowledgement that these products have been produced according to organic production standards. The basic role of a certification body is to confirm that these adhere to a specific organic standard (usually the one established by the importing country or by the certification body itself). Once certified, organic products are marketed, usually carrying a certification label, indicating that the products are certified as organic. The certification label attests to conformity with certain standards and is itself not a trade mark. However, in most countries the certification label is also registered as a trade mark. Producers wishing to enter a specific market for organic products may find it useful to be certified by a certification body with a certification label that is well known in that market.

The cost of certification can be high, although it varies in relation to farm size, volume of production and which certification body is chosen. Relatively few developing countries have certification bodies within their borders (although the situation is changing), and even when sufficient resources are available to pay for certification, farmers often lack the information to find credible inspectors.

To date, government regulations essentially deal with references to the organic production method on labels. They do not define standards for the use of private trade marks or certification marks. This implies that government regulations may be supplemented by private quality seals based on private organic standards.

IMPORT REQUIREMENTS OF THE MAJOR MARKETS

United States

Effective October 2002, any fruit and vegetable labelled or shipped as organic will require certification by an approved certifier based on the US National Organic Standards (NOS). Until then, there is no Government regulation or oversight on organic declarations of fruits and vegetables, which means that imports of organic products are usually undertaken by a United States importer in cooperation with a United States certification body.

For United States organic fresh producers and for exporters targeting the United States market, there are several key factors in the implementation of these standards. All certifiers who apply for USDA certification within the first six months (by August 2001) will have their application reviewed and on compliance, approved by the April 2002 effective date. Beginning in October 2002, the USDA Organic Seal will be allowed on organic products marketed in the United States.

For certification of organic products to be exported to the United States, the exporter has three certification options:

1. US certification bodies operating in foreign countries may apply for USDA accreditation. Foreign applicants will be evaluated based on the same criteria

as domestic certification bodies. In lieu of USDA accreditation, a foreign certification body may:

2. Receive recognition when USDA has determined, upon request of a foreign government, that the foreign certification body's government authority is able to assess and accredit certification bodies as meeting the requirements of the National Organic Standard; or
3. Receive recognition as meeting requirements equivalent to the requirements of the NOS under an equivalency agreement negotiated between the United States and the foreign government.

In practical terms, any group considering exporting organic products to the United States should identify a certification body that has or will receive United States certification approval. The United States based organizations with overseas offices will be able to certify all locations when they are approved by the USDA for organic certification.

Japan

The new Japan Agriculture Standard (JAS) regulations for product labelling formulated by the Ministry of Agriculture Fisheries and Forests (MAFF) have been operational since 1 April 2001. The established JAS law is based on the CODEX guidelines for organic agriculture. Under the new law all products labeled as organic must be certified by a Registered Certification Organization (RCO) and must display the JAS logo, as well as the RCO's name. Under the new regulation, RCOs are required to be accredited by MAFF. Since the start of the implementation of the new legislation, 38 organizations have been registered as RCO. Although it is possible for foreign certifiers to register, at the time of writing, all RCOs are currently Japanese. Standards and regulations governing imports of organic fruit and vegetables into Switzerland are described in the chapter dealing with this country.

European Community (EC)

In the EC, Regulation EEC 2092/91 determines the minimum requirements for organic farming in all the member states and is a directly applicable law. It contains standards for production, processing, imports, inspection and certification, marketing and labelling of organic products. Organic food products originating from non-EC countries may be imported and marketed in the EC with an organic label if it is accepted that the products are produced and certified according to procedures equivalent to those of the EC.

Under Regulation EEC 2092/91, there are, in practice, two options to gain authorization to export organic products to the EC:

1. When a third country has established and implemented organic standards, it may apply to the European Commission for inclusion into the EC List of Article 11. Known as the Article 11 list, countries on this list can export products certified by an approved domestic certification body to the EC without the need for additional certification or accreditation. Note that this is valid only if the domestic certification body is specifically registred in the Article 11 list. Each consignment should be accompanied by a document certifying that standards and certification measures are equivalent to those applicable in the EC.

In order to be included in the list, countries will have to provide sufficient evidence that a reliable system of standards and certification procedures exists, and that it is equivalent to the EC system. The European Commission will evaluate the country's organic standards and its certification measures. These standards should correspond to the standards described in Article 6 of Regulation 2092/91, and certification measures should correspond to the measures described in Articles 8 and 9. Experience shows that it is a difficult and long process for a country to be included into the list. As of June 2001, only 6 countries were on this list (Argentina, Australia, the Czech Republic, Hungary, Israel and Switzerland). For more detailed explanations on how a non-EC country can apply for inclusion in the Article 11 list or how to obtain import permits, readers can refer to the following publication: Organic Food and Beverages, World Supply and Major European Markets, International Trade Centre, Geneva, 1999.

For detailed and up-to-date information on the regulations governing imports and labelling of organic food products into the EC, please see the Web site of the European Commission (http:// europa.eu.int/eur-lex/en/lif/dat/1991/en_391R2092.html). This Web site provides the basic regulation EEC 2092/91 and all its amendments.

2. If the exporting country is not on the Article 11 list, the exporter should ask its importer in the EC country to apply for an individual permit for import. An individual EC member state may authorize an importer to import products from a country not on the Article 11 list. Under this provision, the importer should apply to the designated authorities in the EC member country for an import permission and the application should be accompanied by documentation on the equivalence of standards and control measures. The producer and the exporter will have to be certified in accordance with EC procedures, which has to be done in co- operation with an accredited certification body.

Import permits are issued for a certain amount of specific products from specified countries. They are valid for defined periods. They are granted to specific importers and valid only for those importers for import entry into those importers' country. Once imported into an EC member country, the products may be marketed freely within the EC. In practice, the duration of the process to obtain an import permit can vary considerably. Some importers reported that it is a matter of weeks in some countries (e.g. the Netherlands), while it can take up to several months in other member states. In France, for example, some trade sources said that in the past it used to take up to six months to obtain an import permit. However, they said that there has been considerable progress recently, leading to a more reasonable time frame (generally not exceeding two months).

To obtain an import permit, the importer must provide evidence that the product is produced in accordance with production requirements equivalent to those laid down in article 6, that control measures are as effective as those set out in articles 8 and 9, and that these control measures are effectively and permanently implemented. Equivalency between production methods in the EC and the export country is documented through the use of certification bodies having standards at least equivalent to those of the EC or a certification programme that ensures certification against standards equivalent to those of the EC.

In general, the criteria for granting the import permit are shifting away from evaluation at the production level towards approval of the certification arrangements, including the certification bodies. In order to ensure equivalence in the effectiveness of inspection and certification measures, third-country certification bodies must satisfy the requirements of standards EN 45011 or ISO/IEC Guide 65:1996. To obtain the detailed organic standards applied in major importing countries visit the following Web site: http://www.organic-research.com/Laws and Regs/legislation.htm.

How to Obtain Certification of Organic Fruit and Vegetables in Developing Countries

As explained above, producers and exporters of organic fruit and vegetables seeking to sell their products under the organic label in developed countries have to obtain organic certification. This can be done by the certification bodies of the countries targeted for export, or by other foreign certification bodies, or under a partnership agreement between these two types of certification bodies. To date, relatively few developing countries have certification bodies within their borders, although the situation is changing.

Currently, certification for export in developing countries is often carried out by the certification bodies of the importing countries, even though there are some notable exceptions (e.g. Argentina). The advantage for the exporter is that the logos of these bodies are well known and trusted by consumers in their countries, thus giving the product a better visibility and commercial advantage. The major drawback is that this type of certification can be very expensive, especially when inspectors need to come from the country of the certification body. In order to reduce costs to producers and exporters, international certification bodies tend to use local organic inspectors. Many international certification bodies, such as Ecocert, OCIA or BCS-Öko, have established local branches in developing countries. These branches tend to recruit local staff, which reduces costs. When no local branch exists in a given country, there might be a regional branch in a nearby country that may send a team to carry out inspections and even certification.

Organic certification can also be carried out by local certification bodies in developing countries where these exist, provided that they are accredited by the authorities of the importing country. Experience to date shows that this can be a difficult and long process. To be accredited, a certification body must demonstrate that it complies with basic criteria of transparency and independence (freedom from the influence of vested interests). Certification bodies may be evaluated according to their ability to meet such criteria, and this requires an analysis of their certification systems, including an assessment of their personnel, standards and their inspection and certification procedures. An increasing number of importing countries require that certification bodies comply with the ISO 65 standard or equivalent norms which set out the details of such procedures.

Specific advice on how to establish local certification bodies is provided in the following publication by IFOAM (1997): Building trust in organics: A guide to setting up organic certification programmes, by Gunnar Rundgren.

Because accreditation of certification bodies in developing countries is difficult and therefore rare in practice, local bodies tend to develop partnerships with international certification agencies. In general, the local body carries out the bulk of the activities leading to certification, while the international body periodically evaluates the implementation of certification procedures and sometimes issues the certificates. This may reduce costs for the producers and exporters, while providing access to an international certification mark. It can also strengthen local inspection and certification capacity in the developing country.

OTHER REQUIREMENTS FOR EXPORTING ORGANIC FRUIT AND VEGETABLES

In addition to the above requirements, which are specific to organic products, organic fruit and vegetables destined for export should meet the usual requirements concerning all fresh fruit and vegetables, whether organic or conventional. It is beyond the scope of this study to deal with these requirements, as several publications explain them and provide much detailed and practical information. The following is just a brief summary of the most frequent categories of requirements.

Phytosanitary Certification

Import requirements depend on both the product and the country of origin. In general, a phytosanitary certificate issued by an official of the exporting country must accompany all fruit and vegetable shipments. This official will be able to determine if the particular fruit or vegetable can be exported to the country of destination and what phytosanitary requirements must be met.

Grade and Quality Standards

Fruits and vegetables exported to the EC, Japan or the United States must meet import requirements relating to size, grade, quality and maturity. A certificate based on an inspection must be issued by the country's relevant authority to indicate compliance with standards.

Pesticide and other Contaminants

Most developed countries have established standards for tolerances for pesticides, herbicide and fungicides used in the production and treatment (e.g. fumigation) of agricultural products. These countries have established Maximum Residues Levels, often referred to as MRLs. As organic agriculture prohibits the use of most synthetic pesticides, producers of fresh organic fruit and vegetables are very unlikely to exceed these MRLs. However, those producers who use some of the few synthetic inputs permitted by organic standards should be aware of these requirements.

Import Clearance

In general, the customs services of the importing country are responsible for the final approval and authorization for the importing of all products. Customs services review all documentation for the shipment against requirements and will not release goods from the port of entry until all requirements are met. They are also responsible for the collection of any import duties.

BASIC CONCEPTS OF ORGANIC AGRICULTURE

Accreditation

A procedure by which an authoritative body evaluates and gives a formal recognition that a certification programme is in accordance with the standards of the authoritative body.

Certification

A procedure by which a third party gives written assurance that a product, process or service is in conformity with certain standards. Certified organic food products are food products that have been verified to have been produced in accordance with specified standards for organic production and processing.

Certification Body

An organization performing certification. Sometimes referred to as the certifier or the certification agency.

Certification Label

A label or symbol indicating that compliance with standards has been verified.

Certification Programme

A system of rules, procedures and management for carrying out certification. One certification body may execute several different certification programmes. Sometimes referred to as a certification system.

Competent Authority

The official government agency having jurisdiction.

Control, Control Body

Terms commonly used by the trade when referring to inspection and an inspection body.

Inspection

An on-site visit to verify that the performance of an operation is in accordance with specific standards of a certification programme. The inspector is the person appointed to undertake the inspection.

Inspection Body

The body performing the inspection part of certification. Where a certification body performs its own inspections, the inspection body is identical to the certification body. Sometimes referred to as the inspection agency or the control body.

License

A document issued under the rules of a certification programme, by which a certification body grants a person or body the right to use certificates or certification labels for its products, processes or services in accordance with the rules of the relevant certification programme.

Operator

Anyone carrying out activities covered by a certification programme, for instance farmers, processors, handlers.

Standards

Documented agreements containing technical specifications or other precise criteria to be used consistently as rules, guidelines or definitions, to ensure that materials, products, processes and services are fit for their purpose. Standards relating to organic food products are production and/or processing standards describing, prescribing, allowing or prohibiting procedures and materials, as well as standards for certification and labelling.

Subcontractor

Anyone carrying out activities covered by a certification programme, for instance farmers, processors, handlers, on behalf of an operator.

- EN 45010.1: A European standard for accreditation developed by CEN and CENELEC.
- EN 45011.: A European standard for certification developed by CEN and CENELEC.
- ISO/IEC Guide 61.: An international guide to accreditation developed by the international standardization organizations ISO and IEC.
- ISO/IEC Guide 65.: An international guide to certification developed by ISO and IEC.

The EN 45000 series of standards and their corresponding ISO/IEC Guides lay down general criteria for the operation of testing and calibration laboratories, certification bodies for products, quality systems and personnel, inspection bodies, and accreditation bodies with the aim of ensuring confidence and reliability in the activities of these bodies. The EN 45000 standards are also referred to as harmonized standards.

MAIN MARKETS FOR ORGANIC FRUIT AND VEGETABLES

ORGANIC FARMING IN AUSTRIA

Organic Production

Austria has the highest percentage of organically farmed agricultural land of all the European countries. In 2000 there were about 19 000 certified organic farms, or about 7 percent of all farms. Approximately 272 000 hectares were farmed organically. The average size of organic farms is 14 hectares. Most of the organic production takes place in mountainous areas, where permanent grassland prevails.

The rate of conversion towards organic agriculture in Austria has slowed down in the last two years. While there is a stagnation of organic production, the marketing of organic products continues to increase. Currently, considerable amounts are marketed conventionally. While the increases in production were very high for most products until 1998, the level of production and marketing has remained stable since then. Exceptions are cereals and non- perishable vegetables, to some extent. The rapid increase between 1996 and 1998 was mainly triggered by subsidies. Most farms that benefited from subsidies have converted to organic farming; therefore, further increases will be mainly market driven.

On average, production has more than doubled over the past four years for all product groups. The highest increases are found with pork, but this has to be seen in relation to the very low overall production level. Non-perishable vegetables rank second, while fresh vegetables have shown a rather modest increase.

According to data supplied by VNÖ ("Verein der Naturkostläden Österreichs", an umbrella organization of natural food stores in Austria), the Agrarmarkt Austria (AMA) estimated that Austria experienced an annual growth in organic sales of about 10-15 percent in 1999, among the lowest in Europe.

Governmental Policy

When Austria joined the EC in 1995, producer prices dropped dramatically. Therefore, conversion to organic farming was advocated as the survival strategy for small-scale farming. This contributed significantly to a greater acceptance of organic farming within the agricultural system. The Austrian Government had already started to give subsidies for conversion to organic agriculture in 1992. In 1995 a horizontal agro-environmental programme under EC Regulation 2078/92 was effected. The programme, called Österreichisches Programm für umweltgerechte Landwirtschaft (Öpul), was slightly adjusted in the year 2000 and extended for five years. The payments per ha increased and are presently as follows:

Table. Subsidies for Organic Production Under the Agro-Environmental Programme (OPUL)

Category	1998	2000
Arable land	S4 000/ha	3 450
Meadows	S3 000/ha	7 000 to 9 000
Vegetables, strawberries	S6 000/ha	11 000
Horticulture, tree nurseries, hops	S10 000/ha	4 500

Table. Total Organic Production of Fruit and Vegetables 1999

Product (tonnes)	Total	Indirect	Direct	production
marketing	marketing			
Potatoes	12 000	10 000	2 000	
Other storable vegetables	6 000	5 000	1 000	
(carrots, onions, etc.)				
Fresh vegetables	3 000	2 500	500	(tomatoes, green
pepper, lettuce, cucumber, etc.)				
Total vegetables	21 000	17 500	3 500	
Total fruit	6 000	5 000	1 000	

Virtually no official statistics are available for the organic sector on fruit and vegetables in Austria. This underlines the minor importance of this sector within Austrian agriculture. Import data for organic and conventional products are not separated. Market partners and importers were reluctant to supply hard data. The Austrian market of organic products in general is dominated by supermarket chains, which are in the process of establishing their own organic labels. They were disinclined to release any data, as they fear loss of competitive advantages. Therefore, in order to at least get a general picture of the situation, the author has made estimations and rough calculations based on estimates. These estimations were cross-checked with market experts and seem to be realistic. However, the results have to be seen in the light of these limitations.

THE AUSTRIAN MARKET FOR ORGANIC FRUIT AND VEGETABLES

The Organic Market

In 1999 the total organic turnover was estimated at S2.9 billion, or 1.8 percent of the total food market. The annual growth rate is estimated at 10-15 percent. About 70 percent of the turnover is made in urban areas.

Table. Organic Production Marketed as Organic 1996-1999 (Indirect Marketing Channels Only)

Product	1996	1997	1998	1999	Average Growth Rate/Year
Beef (numbers)	3500	6200	7800	8000	32.14
Pork (numbers)	1000	5000	10000	10000	225
Poultry (numbers)	150000	160000	200000	200000	8.33
Milk (mill. litres)	75	100	135	150	25
Eggs	500000	1500000	1600000	1800000	65
Cereals (tonnes)	16000	22000	28000	38000	34.4
Potatoes (tonnes)	5500	9500	13000	13000	34.1
Other less perishable onions)	1000	2600	5000	7000	150 vegetables (carrots,
(tonnes)					
Perishable vegetables	2000	2000	2500	2500	6.25 (tomatoes, green
pepper,					
lettuce, cucumber, etc.)					
(tonnes)					
Total vegetables (tonnes)	8500	14100	20500	22500	41.2
Total fruit (tonnes)	1120	2150	1500	1200	1.7

VNO expects that the share of organic products within the total turnover in the food sector will rise to more than ten percent (compared to the present two percent) within the next ten years (VNO, 2000). Although this might be an optimistic statement, there is potential for a steady market-driven growth of the organic sector.

Sales of Organic Fruit and Vegetables

About three percent of all fresh fruit and five percent of all vegetables sold in Austria are produced under certified organic production standards (RICHTER et al, 2000). Table provides estimates of sales value using these percentages. The table shows that sales value of organic vegetables is estimated at S61 million, while organic fruit sales are estimated at S170 million. The value found in literature is S200 million (RICHTER et al, 2000).

Table. Total Sales Value of Fruit and Vegetables and Estimated Organic Sales Value (in 1 000 Austrian Shillings)

Total sales value				Estimated portion organic
	1999	**2000**	**1999**	**2000**
Fresh vegetable	5 011 543	5 230 472	250 577	261 523
Fresh fruit	5 629 830	5 663 402	168 894	169 902
Total fresh fruit	10 641 373	10 893 874	419 471	431 425 and vegetables

Average Prices at Retail Level and Premiums Over Conventional Products

No statistical data is available for prices of organic products. Therefore, the authors had to rely on their own assessment, based on supermarket observations.

Table: Assessment of Premiums for Organic (January 2001)

Product Conventional	Price organic (percent)	Price	Premium
Potatoes	12.90/kg	8.90	45
Carrots	10.90/kg	4.90	122
Onion	14.90–/kg	7.90	89
Broccoli	12.90 – 15.90/kg	10.90	18- 45
Oranges (Italy)	19.90/kg	14.90	33
Grapefruit	9.90/piece	7.90	25
Lemon	16.90/kg	12.90	31
Kiwi	2.90/piece	1.66	74-135
Apple	24.90/kg	19.90	25
Pears	29.90/kg	24.90	20
Tomatoes	24.90/kg	19.90	25

The range of premiums for organic fruit and vegetables at the supermarket level s, on average, around 20-30 percent, in some cases they are much higher (for example, carrots). This corresponds well with the range of premiums found in literature on organic horticultural products. The assessment varies, however, according to price specials and can therefore be regarded only as indicative. Premium levels differ, depending on region as well as on the type of distribution channel. In natural food stores the prices for organic products are generally higher. Premiums paid to organic producers are estimated by market experts between 30-100 percent.

Distribution Channels

The main organic sales outlet is a supermarket. A sector analysis by the Austrian umbrella organization for natural food stores (VNO) shows that 66 percent of organic products are marketed by supermarkets, 22 percent by natural food stores and about 12 percent by direct marketing.

For sales of organic fruits and vegetables these percentages differ, as shown in Table. Direct marketing through farm sales is the main sales channel for fruit and vegetables.

Natural food stores cooperate with smaller farmers' associations, which are organized under the umbrella organization OIG (Osterreichische Interessensgemeinschaft für biologischen Landbau). Marketing into the supermarket chains is organized by the farmers' association Ernte für das Leben, which has a daughter firm called "Ökoland". This firm collects the products for supply to supermarkets and also for export.

Table. Distribution Channels for Organic Fruits and Vegetables 1996

Direct sale of	Delivering services	Natural food	Supermarket farmers of farmers' initiatives stores
49	9	19	23

The supermarket chain "Billa/Merkur" launched the organic food brand "Ja!natürlich" in 1994. All major supermarket chains (including discounters), such as Adeg, M-Preis, Maximarkt and Spar ("Natur pur" dairy products) now offer organic products.

Within the supermarkets, "Billa/Merkur" with the "ja!natürlich" label (www. janatuerlich.at), holds about 70 percent of the organic market share, followed by the supermarket chain "Spar", with 20 percent. Spar has an organic label called "Natur pur". The two supermarket chains differ in their policy towards organic products. While "Billa" has a strong, central unit dealing with organic products only, Spar's organic products fall under the responsibility of each separate product department.

Austria has about 150 natural food stores, many of which are very small. Recently an umbrella organization called "Verein Naturkostläden Österreichs" (VNÖ) was formed. About 33 natural food stores are members, most of which are small with only 3 shops that have more than S3 million annual turnover. They are concentrated in eastern and southern Austria.

The natural food stores fill the gap between supermarkets and farmers' shops and farmers' markets, which remain an important distribution channel for organic products.

Trends and Attitudes

A recent trend can be recognized towards the creation of "organic supermarkets", with a full range of products and more than 300 m² of floor space. The first organic supermarket opened in 1999 in Vienna. The strength of this distribution channel lies in the fact that it combines the convenience of ordinary supermarkets with the advantages of natural food stores (specialized in the organic sector and more detailed consumer information).

A relatively new but growing segment is the provision of organic products in public canteens. Recent food scares motivated Government officials and labour unions to put more pressure on public canteens to incorporate a higher share of organics into their purchases.

According to market research reviewed, Austrian consumers in general still prefer fresh products to convenience products and rank health higher than time-saving. Regarding quality criteria, "freshness" is ranked first (68 percent), followed by "natural" (27 percent) and "origin" (20 percent). Austrian origin is regarded important by 19 percent. (BMLF, 1997).

For horticultural products, the trend of sales since 1996 shows a varying picture. While the share of production marketed as organic potatoes and other storable vegetables has increased, fresh (and perishable) vegetables and fruit sales have been almost stagnant. Only a slightly positive trend is expected in the long run by market experts. The market seems to be too fragile, and therefore, market partners are unwilling to take a risk with perishable products.

Table. Annual Growth Rate of Marketed Production in Organic Horticulture 1996- 1999

Product (tonnes) rate/	1996 year	1999 (percent)	Average growth
Potatoes	5500	13000	45
Other storable vegetables	1000	7000	200 (carrots etc.)
Fresh vegetables (tomatoes, green pepper, lettuce etc.)	2000	2500	8.3
Total vegetables	8500	22500	55
Total Fruits	1120	1200	2.4

Based on the market research reviewed, 54 percent of Austrians at least occasionally buy organic products. The typical consumer of organic products buys fruit and vegetables (79 percent), milk and milk products (61 percent) and cereals (52 percent). Consumers of organic products are usually those with higher income and higher education in the urban population.

The organic "fundamentalists", those who strictly buy organic products, still favour direct sales in farmers' markets, on-farm sales or natural food stores. This group is becoming smaller, however, while the group of occasional buyers is growing. Since all major supermarkets offer a share of organic products, the turnover of organic sales has increased considerably.

Constraints to Market Development

Constraints to market development are found on production as well as marketing sides. On the production side, due to climatic constraints, Austria is limited to a certain range of products. Most farmers producing organic vegetables are small-scale farmers who have a considerable history in direct sales. In the fruit and vegetable sector direct sales still play a major role. Production of small-scale farms cannot be increased to meet the requirements of supermarkets. Conversion of large-scale farms is difficult as their specialization prevents the nutrient cycling and pest management required in organic farming. Therefore, a high percentage of the fruit and vegetables (conventionally and organically produced) needs to be imported.

On the marketing side, consumer attitudes and price expectations are the strongest limitations. Austrian consumers prefer regional products. Freshness is very important for buying, especially in the vegetable sector. For some consumers (especially for the ecologically conscious) ecological transport is relevant since transit, especially in the Alpine regions, is a major issue of the ecological debate. For organic products, the confidence in certification mechanisms is important. The highest confidence is reached through direct sales by the producer. Direct sales further reduce exact price comparisons. Most supermarkets offer organic products under their own labels. Trust in the labels of supermarket chains is less than in the labels of producers. Producer labels are found more frequently in natural food stores and on-farm shops.

Sales via supermarkets has increasingly reached new consumers' groups, but as they offer organic and conventional products side by side, the price becomes the main criterion of choice.

IMPORTS OF CERTIFIED ORGANIC FRUIT AND VEGETABLES INTO AUSTRIA

Market Access

EC Regulations regarding market access are applied. Besides the Regulation on organic production, there are market regulations for fruits and vegetables and processed products in general. Within the frame of these market regulations, the common norms (trade classifications) for fruit and vegetables are followed. Imports and exports of fruit and vegetables are furthermore restricted by licences and quotas. Licences are needed amongst others for the import of fresh garlic from China, tomatoes from Morocco, and for processed fruits and vegetables according to EC rule (EG) Nr. 1921/95.

The import volume of specific varieties of fruits and vegetables is monitored by an online system, whereby customs offices transfer data on imported quantities to the European Commission on a weekly basis (AMA; www.ama.at). Besides these general trade restrictions, the EC has special norms for organic products. In addition to the EC Regulation, all market partners in Austria require certification by an Austrian certification body (e.g. "Austria Bio Garantie") to be sure of adequate standards.

Total Import of Organic Products

No statistical data is available on import totals of organic products. Most key sources estimated the value of imports at S1 billion. Considering a total turnover of organic products of S3 billion, this would come to about 30 percent. Also, experts answering in a study on supermarkets (RICHTER et al, 2000) assume that about 30 percent of the organic products are imported. The main imports come from the Netherlands, France, Germany and Italy (CIR, 1999, p 38). The highest proportion of imports is fruit and vegetables.

However, if converted into absolute numbers, these figures do not match those given for overall imports in the organic sector. As already mentioned, the total value of imports is estimated to be S1 billion. The total value of the organic fruit and vegetable sector is estimated to be about S430 million. The total value of vegetables was estimated at S261 million. A 50 percent import share would mean about S130 million. The authors' assessment after discussions with market participants regarding organic vegetable imports resulted in about S169 million. For the fruit sector, the authors' estimates resulted in a rough figure of S158 million. This would total up to S327 million.

Therefore, considerable amounts of other products would need to be imported to reach that total. The amount of cereals imported, totals 2 000 tonnes (BMLF, 2000). Even under the assumption that the 20 percent (mentioned in Table) would total up to 7 000 tonnes, this still cannot make up for the difference. It can only be assumed that there are a number of smaller items, such as processed products, which in total might add up to the reported value. On the other hand, it is also possible that the import value is greatly overestimated.

Total Import of Organic Fruit and Vegetables

Market sources estimate that between 1-5 percent of total imports of fruit and vegetables are organic. Since 1999 total import value of fresh fruit was S5.4 billion, while fresh vegetables were valued at S3.2 billion (total value for imported fresh fruit and vegetable was S8.6 billion); organic imports are estimated to range between S86 million and S370 million. According to the report of the Corporate Intelligence on Retailing (CIR, 1999), about 70 percent of organic fruit and 40 percent of organic vegetables are imported. Given the figures for the sales value of organic fruit (S170 million) and vegetables (S261.5 million), this would mean a value of S104.6 million for vegetables and S119 million for fruit. The total import value thus would be S223.6 million.

The authors' own assessment of major fruit and vegetable imports resulting from interviews with market participants gave a total of S176 million for vegetables and S158 million for fruits. This would total to S334 million. Therefore, it can be assumed that the total import value is around S300 million.

Breakdown of Organic Fruit and Vegetable Imports by Product

According to market experts, there are two rules of thumb to be applied on the situation of imported organic fruit and vegetables:

- The more perishable an organic product is, the less interested importers and retailers are. While this attitude is general and not restricted to the organic sector, the fact that the organic market is still weak and that retailers cannot rely on known consumption patterns, contributes to this situation. Non-traditional products (like exotic fruit and vegetables) are especially perceived to have a higher commercial risk.

- The premium for organic products over conventional products is a critical factor. This is true for all organic products. However, it applies especially to imported organic products as trust in foreign certification is low and organic products are easily substituted by conventional products if there is a high premium. In the long term it is expected that premiums will decrease, especially in supermarkets.

These restrictions influence, for instance, the situation with bananas. On the one hand they do not store very long, but on the other hand, their premiums are the highest within the group of tropical fruits. Off-season organic berries (strawberries, etc.) and other perishable fruit are not imported in substantial amounts as the market is not yet stable enough and the risk of losses is high. Exotic fruit (mango, papaya, etc.) fall into a similar category. There is no organic demand for these fruits. They are highly priced luxuries and do not constitute a substantial share of the daily consumer purchases.

Table. The Main Imported Organic Vegetables

Product	Quantities (tonnes)	Value (ATS 1000)	
Zucchini	2 700	67 500	
Kohlrabi	2 500	17 500	
Broccoli	1 000	17 000	
Potatoes	900	7 000	
Onions	700	7 000	
Carrots	1 200	18 000	
Fennel	300	7 500	
Tomatoes	100	2 000	
Others (green pepper,		32 500	cherry tomatoes etc.)
Total	9 400	176 000	

As for organic fruits, the main products sold are oranges, followed by lemon, apples and kiwi. The results are given in Table.

Table. The Main Imported Organic Fruits

Product	Quantities (tonnes)	Value (ATS 1000)
Oranges	4 000	65 000
Lemon	3 400	50 000
Kiwi	1 000	20 000
Apple	400	8 000
Others (bananas, pears,		15 000 mango, avocado)
Total	8 800	158 000

Trend for Each Product

Strong increases have been seen for organic zucchini and fennel. However, imports of organic tropical fruit is growing slowly, as particular organic quality is not requested, and the total quantity is low. Market sources expect a rather strong positive trend for citrus and carrots (as the domestic supply cannot meet the growing demand), especially in the baby food sector.

Main Importers

Main import structures for organic fruits and vegetables include direct imports by supermarkets, general import wholesalers on contract with supermarkets, specialized importers for organic products and specialized importers for fruit and vegetables.

Supermarkets

Some of the big chains, like "Spar", import themselves. Presently, they import organic fruit and vegetables from producer cooperatives in Mediterranean countries. "Billa" imports through wholesalers who are also in the conventional fruit and vegetable business. It seems, however, as if they are only involved with logistics, since all organic farmers supplying "Billa" (also those outside Austria) need a contract with the label "ja!natürlich". "Billa" does not only issue a frame contract but also negotiates price and quality. "Billa" usually issues long term contracts to potential producer groups for the most important products. This is true for Austrian producers and for producer groups in other countries

Specialized Importers of Organic Products

Supply other distribution channels (natural food stores, specialized fruit and vegetable shops, small processors in gastronomy, etc.). Most operate either via Austrian daughter firms of German importers (e.g. "AL Naturkost" in eastern Austria), or they operate directly from Germany (Dennre, Bodan), as the Austrian market is comparably small and as "AL Naturkost" does not import directly from the exporting country but via importers in Germany.

Conventional Importers of Fruits and Vegetables

Have a very limited share of organic products so far. They have a relatively stronger position in western Austria, while organic specialists especially cover Vienna and eastern Austria. There is a growing interest among some conventional importers to move towards dealing with organic products, but one major constraint they raise is the strong market position that is built up by "Billa", contracting with only a few importers. Others: some organic food processors need components (such as potato starch or white sugar) in limited amounts. " Bioservice" (a daughter firm of the biggest organic farmers' association "Ernte für das Leben") specializes in importing such products. "Hipp", a major processor, is the largest producer of organic baby food. For vegetables there is a specialized firm that deals with domestic, as well as imported, products (carrots, onions, potatoes, etc.). According to "Hipp Austria" (Gmunden), tropical ingredients are imported for the central firm in Pfaffenhofen/Germany and the Austrian firm mainly receives final products. Another processor, specialized in dried fruit and mixtures with cereals (for muesli), is "Naturprodukte Perlinger". It imports and exports dried fruit (partly only re-labelling, with distribution via supermarkets).

Product Specification

Products are only marketable if they comply with the requirements of the trade classification in categories "Extra, I or II". The classification system acts according to the Austrian federal law on quality classes.

According to importers, the reduction of storage life due to the prohibition on use of storage chemicals does not affect organic products in their sector. This is a major problem, however, on the retail level. Some retailers try to increase shelf life of sensitive organic products by packaging organic fruits and vegetables in cardboard boxes with plastic wrap, but that contradicts the consumers' expectations of "natural" and "freshness".

Prices at Import Level

AMA gives import prices for organic and conventional products. Prices vary considerably according to the country of origin. While some importers claim that prices are almost the same as for conventional products, others claim that a 10-15 percent premium is charged for oranges and lemons. According to supermarket chains, oranges and lemons from Italy (the main organic products) achieve a premium of up to 30 percent. The difference might be explained by the fact that some supermarket chains have their regular suppliers, while others buy on the daily market.

Table. Prices and Premiums for Selected Organic

Product origin	Price conventional	Country of origin	Price %	Country of	Premium (ATS)	organic (ATS)
Tomatoes	31.5/kg	Italy	25/kg	Spain	26	
Lettuce	133/kolli	France	122/kolli	Italy	9	
Green	34.65/kg	Italy	34/kg	Spain	2	pepper
Onions	17.15/kg	Italy	18/kg	Italy	–	
Apples	20/kg	Germany	16/kg	Italy	25	
Oranges	19/kg	Spain	19/kg	Spain	–	
Lemon	18/kg	Spain	18/kg	Spain	–	
Bananas	26.95/kg	Dom. Rep.	22/kg	Panama	22.5	

An assessment was made, comparing price lists of German wholesalers specialized in organic products, which also export to Austria, with the price list of an Austrian importer of conventional fruit and vegetables.

According to this assessment, the premiums for citrus seem to be practically nil. Tomatoes fetch a rather high premium. However, because of storage problems with organic tomatoes, the amount of imported tomatoes is low.

Relatively high premiums are paid for bananas, however market experts attribute this to the small organic banana market. The low market share and the short shelf life result in the high premiums at the import level (according to one importer, it can reach up to 100 percent at times), which is a sort of risk payment. On the other hand, the high premiums restrict market development.

Major Suppliers and Main Products

The Austrian market for organic fruit and vegetables relies mainly on suppliers from the European Community. One reason is because retailers (and consumers) do not trust certification in overseas countries. Within the EC, no import restrictions exist. Therefore, European countries are the main sources of origin.

However, sometimes consumers are reluctant to buy non-Austrian products. As some countries are regarded as practising intensive agriculture, many organic consumers do not buy organic products originating from them. The main suppliers of fruits and vegetables are in southern Europe or in Mediterranean countries.

Table shows a variety of countries of origin, for which data has been taken from lists of importers and retailers showing the country of origin. It has to be kept in mind, however, that the quantities for most of these products are very low.

The main import season for many products is the off-season (winter). Potatoes are imported only in spring, especially for the asparagus season (which are eaten traditionally with boiled potatoes) when fresh, but domestically produced potatoes are not available and the old stock is no longer of sufficient quality.

Table. Main Countries of Origin for Imported Organic Fruit and Vegetables, Major Products, Major Season of Supply

Country	Product	Season of supply
Israel	Green pepper, Chinese cabbage	Winter
	Potatoes	Spring
	Avocado	All year
	Tomatoes	February to May
	Oranges, grapefruits, tangerines	All year
Italy	Apples	Winter
	Oranges	October to March
	Lemons	All year
	Fennel, aubergines, green pepper,	Winter
	Tomatoes, lettuce	
Spain	Oranges, lemons	October to March
	Grapes	Winter/spring
	Cucumbers, cabbage, broccoli	Winter
	Sweet potatoes	All year
France	Cabbage, lettuce	Winter
	Garlic, shallots	
	Parsley	
Greece	Oranges,	October - March
	Lemons	All year
Dominican	Bananas	All year Republic

Constraints to Import Growth

Major marketing institutions (like the AMA) and the general agricultural policy inspire the general consumer to "buy Austrian". Many consumers consider Austrian products more natural and better than those from abroad. This idea is further fostered by the recent food scares in the meat sector (BSE and foot and mouth disease).

The main constraint to import growth in the fruit and vegetable sector is the lack of confidence in the certification bodies of non-EC countries. A second constraint comes from organic farming associations (especially their biggest organization "Ernte für das Leben"), which are involved in the main organic labels of supermarket chains. Usually their market representative are consulted before new products are listed. They try to argue against products that can be grown in Austria or that substitute Austrian products (like organic apples from Argentina), and in general, retailers share this view. The third constraint is that consumers are unwilling to pay the premiums for organic products, especially at supermarkets, where consumers can choose between organic and conventional products.

SUMMARY OF KEY CHARACTERISTICS OF THE MARKET

More than half of the organic fruit and vegetables on the Austrian market are sold directly by farmers. Supermarkets sell about 23 percent of all vegetables, and natural food stores sell about 19 percent. Imported organic fruit and vegetables are sold either by supermarket chains (about 55 percent) or by natural food stores (about 45 percent).

There is a significant demand for organic fruit and vegetables that cannot be produced domestically, mainly due to climatic conditions. However, consumers as well as retailers prefer domestically produced organic products, and only purchase imported products if they are not available in Austria. Imports from Mediterranean countries are preferred over imports from countries farther away. Retailers fear produce loss due to limited shelf life, and therefore, the highest imports are seen for those products that are not so easily perishable.

Main Opportunities and Constraints

Opportunities for foreign exporters to supply the Austrian market are available because Austrian production cannot meet demand. Therefore, especially during wintertime, a range of fruit and vegetables has to be imported, including fruits, such as oranges, lemon, apples, kiwi, bananas, as well as off-season vegetables, among which zucchini, kohlrabi, broccoli and fennel are the leading products.

However, retailers give priority to imports from producers in neighbouring countries, reducing prospects for developing countries to export organic products to the Austrian market. Exotic fruit and vegetables, such as papaya, mango or sweet potatoes are not demanded in large quantities and usually not in organic form. Retailers are mostly interested in products that store well. Apples come mainly from South Tyrol in Italy, a region with strong cultural ties to Austria.

There are severe constraints to imports from developing countries:

- Consumer perceive credibility of organic standards abroad as low. Prospective exporters need to ensure that the certification is comparable to European standards;
- The biggest organic farmers' association (Ernte) is a partner in almost all large-scale marketing projects and plays a key role in market access. It tries to obstruct products that might compete with domestic production (like apples or potatoes), unless off- season;
- Typically, organic tropical fruit is not so much in demand; and
- Retailers are reluctant to stock products with short shelf life. This applies especially to organic products, as retailers are unsure of the expected demand.

Entry Strategy and Recommendations for Export Development EC registration is a prerequisite for market entrance. Generally there are two possible strategies for market entrance into Austria:

- Directly via supermarket chains, which have long-term contracts with exporting producer groups, or via wholesalers and importers. Retailers request the producer group in the exporting country to adhere to the international standards for organic production. In order to build up trustful relations, most retailers, especially supermarket chains, request the approval of a national Austrian certification body, like ABG. They, in turn, are affiliated to organic growers' associations, which are resistant to imports that might substitute domestic production.
- Via specialized import firms that supply the natural food store sector. In this case, the Austrian market should not be seen independently from the bigger German market, as this is the traditional partner for imports of organic products. Many of the importing firms specialized in organic products either operate directly from Germany or are Austrian daughters of German firms. Entrance into the Austrian market thus, to a large extent, is via Germany. Also, natural food stores are linked to organic farmers' associations, which are sensitive to products that might substitute domestic production.

Within the organic movement however, there is a strong sympathy for "fair trade" products. According to a recent study, 32 percent of Austrian consumers are interested in fair trade product. (Hortner, 2001). Chances seem to be best for organic combined with "fair trade". This means that the rules for production do not only include organic standards but also social aspects. Austrian certification bodies apply reduced rates for certification of farmers' groups in developing countries, if involved in organic "fair trade" projects. The long-term trade relations, characteristic of the "fair trade" movements, are also found in contracts issued by Austrian supermarket chains. Therefore, the most promising entrance strategy seems to be close cooperation with Austrian organic growers' associations (and certification bodies), as well as with "fair trade" organizations.

❐

12.

Ailments of Leaf

A number of leaf spotting or blotching diseases occur on trees and shrubs. Some of these are severe to plant health or aesthetics, such as rose black spot and scab on crabapple, but many are of minor effect. Following are descriptions of several of the more common leaf spot and leaf blotch diseases.

PLASMOPARA VIBURNI

This is one of the relatively few significant downy mildew diseases of ornamentals. Spores produced in dead plant tissue that overwinters splashes to foliage in the new season, causing patchy lesions on upper leaf surfaces and downy grayish-white fungal growth on lower leaf surfaces corresponding to these lesions.

Damage starts out with light greenish spots that grow together forming angular patches often bordered by veins. Leaf tissue in the blotches often reddens, browns, dries, shrivels and leaf drop sometimes occurs. The fungus reproduces rapidly and multiple infections occur during periods of leaf wetness and cool to warm, but not hot conditions.

Keep leaves as dry as possible by avoiding overhead irrigation and enhancing good air movement by proper plant siting and pruning practices. When applying preventive fungicides make sure to get good coverage of the lower leaf surfaces where infections occur. When fungicides are required, use a labeled product containing mancozeb. This disease is diagnostically different from powdery mildew of viburnum in that the fungus is present on lower rather than upper leaf surfaces.

BOTRYOSPHAERIA OBTUSA

This leafspot disease is a minor problem on most crabapples, although a few exhibit considerable leaf yellowing and leaf drop in some locations, including 'Madonna' and 'Professor Sprenger.' During moist spring weather the fungus infects leaf tissue and causes small roundish brown spots with purple borders. In some cases spots enlarge developing irregular lobes that, when they enclose the original roundish spots, cause a "frogeye" symptom.

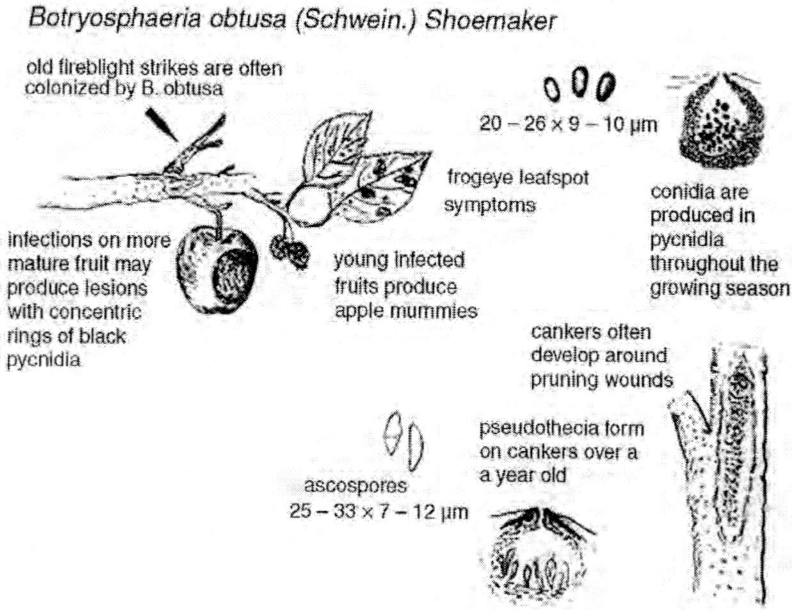

Fig. Life Cycle of Botryosphaeria obtusa

In addition to the leaf yellowing and defoliation that can occur on particularly susceptible cultivars, black rot cankers can develop on stems over the years, especially on crabapples weakened by winter injury and other environmental stresses. In the rare cases when fungicides are warranted, begin sprays at budbreak. When fungicides are used, apply labeled products containing mancozeb or chlorothalonil.

TAPHRINA CAERULESCENS

This is a minor disease in the northern states, occurring on both the red oak and white oak groups. Symptoms include light green to yellowish blister- like, roughly circular bulges on upper leaf surfaces (depressions as viewed from the lower leaf surface). These blisters tend to brown as the season progresses. Spores overwinter on leaf buds and infect leaves as they open. This is the only infection that occurs each season, and leaves become resistant to infection as they mature. Mild, moist spring conditions favour infections. Controls are generally not warranted and fungicide recommendations are rarely made except in certain nursery/garden centre situations in which fungicides are applied dormantly before leaves emerge in the spring. When fungicides are required, use a labeled fungicide containing chlorothalonil, mancozeb or thiophanate-methyl.

GUIGNARDIA AESCULI

This disease is a serious aesthetic problem on most horse chestnut and buckeye species, although bottlebrush buckeye (Aesculus parvifolia) exhibits excellent resistance.

Large irregular reddish-brown lesions with surrounding yellowed tissue occur on leaves, often badly disfiguring foliage by early to mid summer.

Leaves often curl and brown and, by August, the overall plant often looks as if it was blow-torched. Early leaf drop also occurs. The problem is enhanced by wet foliage conditions and the disease is not a problem in drier sites.

Initial infections are in spring from spores produced in infected leaves from the past year. Moist conditions enhance the infections and subsequent cycles of infection occur if moist conditions continue. Black fruiting bodies of the fungus are often evident in lesions.

The disease does not appear to be a serious health problem, as much of the annual growth of Aesculus has occurred by the time foliage is badly damaged. Controls for the serious aesthetic damage include fungicide applications made as leaves emerge, with repeated applications at 10 to 14 day intervals if wet conditions persist.

Use a labeled fungicide containing chlorothalonil or mancozeb. Also practice sanitation by cleaning up infested foliage at the end of the season, and improve air movement in the tree canopies to hasten leaf drying.

RHYTISMA ACERINUM, RHYTISMA PUNCTATUM

These dramatic but inconsequential diseases of numerous maple species cause small to almost one-inch diameter tar-like spots on leaves. The fungus overwinters on fallen leaves, then infects the upper surfaces of leaves in spring during moist conditions.

Leaf spots are first a yellowish green but by mid to late summer a tar-like mesh of fungal and leaf tissue develops inside the yellowed area. Occasionally some leaf withering and drop occurs but this is not generally serious and fungicide sprays are not generally recommended. If fungicides are required, use a labeled product containing mancozeb or triadimefon.

PHYLLOSTICTA MINIMA

Like many fungal leaf spots, this disease affects a number of maple species, most prominently Amur, Japanese, red and silver maple. The disease causes little damage because the infection is localized. Spots are roughly circular and develop into tannish spots with purple to red borders.

Later in the season the spots often contain black fruiting bodies of the fungus arranged in rings inside the lesion. Although this disease is quite noticeable in the landscape, especially on silver and red maples, and causes concern among homeowners, damage is minimal and fungicides are rarely necessary. If fungicides are required use a labeled product containing mancozeb or chlorothalonil.

Some common product names containing the fungicides mentioned in this fact sheet include: mancozeb (Fore, Dithane, Mancozeb); chlorothalonil; Thiophonate-methyl and triadimefon (Bayleton, Strike). It is the user's responsibility to be certain that the fungicide being used is labeled for the specific plant being treated.

CARE FOR TREE WOUNDS

One of the most common types of tree disease is wood or trunk decay. Decay in a tree can result in dangerously weak trees, unsightly trees, or in shortening the life of the tree. Decay in a tree cannot be cured. Proper tree care can, however, prevent decay or greatly limit the progress of decay in the tree.

Decay is a condition that results from the digestion of wood by fungi and other microbes. Wounds through the bark start the processes that lead to decay. Wounds that expose the wood can be caused by animals, insects, birds, fire, storms, freezes, and the activities of humans, such as pruning. Of course, not all wounds lead to decay. Proper care of tree wounds can lessen the chance that decay will result. When a tree is wounded, a natural process begins that results in the covering of the wound by bark and new wood. This is called wound closure, "callusing over," or "wound healing." Natural processes also begin that make wood beneath the wound unsuitable for growth of decay organisms. Proper treatment of wounds can hasten both of these processes.

Proper care of tree wounds begins with practices that promote callusing growth. Recent research work has shown that when a tree produces callus rapidly, the extent of the natural discolouration of the wood in the trunk above and below a wound is less than when callusing is poor. This is important because decay takes place only in discoloured wood. Rapid callus production commonly occurs in the spring when trees are growing vigourously.

Thus, to reduce decay hazard, the best time to make large pruning wounds is from the time that buds start to swell until the leaves are nearly fully expanded. This is usually in March, April, or May for much. The sort of large wounds we are referring to are those involving branches larger than 11/2 inches in diameter.

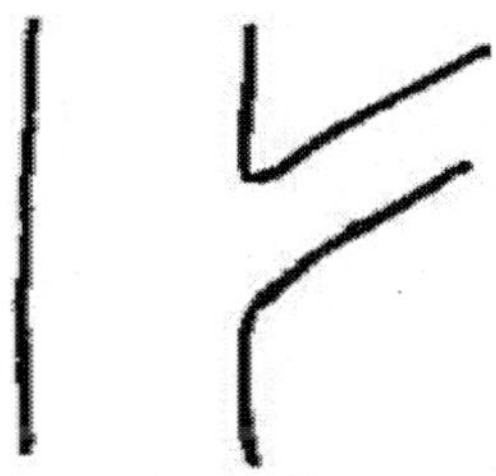

Fig. Trim Branches without Collars Flush to the Trunk.

In the past it hasbeen recommended that the bark around ragged wounds on the trunk or large branches be trimmed away from the wound in an elliptical form, with the greatest dimension of the ellipse above and below the wound. Recent research suggests that when wounds are made in the spring (as recommended above) this is not a good practice.

This is because trimming away healthy bark may lead to increased wood discolouration and hence to increased decay. Elliptical trimming may not be important when wounds occur in late summer or fall either. The bark over these sorts of wounds tends to die and become loose in an elliptical fashion naturally. Healing associated with branch stubs may often be delayed and lead to increased decay within the tree if the stub is not properly pruned off. Cutting flush with the tree is generally best. Some branches have collars, "shoulders" or other natural projections.

These should not be trimmed off. When natural projections or collars are removed, not only is there a larger area to be callused over, but chances of decay organisms entering the wound and becoming established are greater. Branch stubs previously incorrectly pruned should be corrected as promptly as possible. Care should be taken, however, not to cut into the new callus tissue that has begun to form in response to the wound.

Try to keep wounded trees growing vigourously. This will increase the rate of wound closure and reduce the amount of discoloured wood and decay associated with wounds.

Trees should be fertilized properly and watered during droughts. Aerate the soil if it is compacted. Do not plant trees that will grow to be large too close to a house, paved area, or other restrictive object or soil situation.

Thin out less valuable, competing trees. Take measures to control debilitating insect infestations. Protect the tree from wounds made by careless mowing or weeding practices. Prune badly broken branches promptly. You may use wound dressings if you wish; however, some tree care specialists feel that wound dressings interfere with healing or callusing. A light application of tree wound dressing does indicate that someone attended to the problem and may make a pruning job look better or more professional. Trees in the urban landscape must be helped. Early in the life of the tree, establish sound maintenance programmes for all trees. Part of such maintenance is the prevention of wounds, maintenance of tree vigour, and proper wound care.

SOOTY MOLDS ON TREES AND SHRUBS

Symptoms

A sooty, gray-black, velvety, often crust-like coating may develop on the leaves or needles, fruits, and branches of certain plants. The coating is actually the growth of one of several species of black-coloured fungi or molds. The coating can be removed easily by rubbing the leaf between the fingers, thus exposing the green leaf tissue below. Sooty molds grow only on the plant surface and will not kill plants. In fact, sooty molds often grow on sidewalks or fences under infested trees. Sooty molds are normally considered to be a cosmetic or aesthetic problem.

In extremely severe cases, it is possible for the black growth to block enough sunlight to interfere with photosynthesis. In such cases, leaves or needles, fruits and new shoots may be smaller, or less intensely coloured. Respiration can be reduced through the physical closure of stomates by the molds' vegetative growth. Under drought conditions, plants affected with sooty mold will wilt more rapidly than unaffected plants. If plant vigour has been reduced, the plant may also be predisposed to further injury by other insects, diseases or environmental stresses.

Insect Association

Sucking insects are the primary cause of sooty mold growth. Many plant sap-sucking insects feed on leaves and stems of trees and shrubs. These sucking insects often produce excessive, watery excrement rich in sugars. This excrement is called honeydew. Excreted honeydew often falls on leaves or needles, branches, fruits or anything else immediately underneath the infested area of the plant. It is on this excretion that the sooty mold fungi grow. Sometimes plants not actually infested by insects may be affected if a tree above them is being attacked by a honeydew producing insect and the honeydew drops onto them.The major types of honeydew producing insects are: Aphids, soft scales, mealybugs, whiteflies, leafhoppers, planthoppers and psyllids. Occasionally, the spittle-like froth produced by spittlebugs promotes the growth of sooty molds.

Susceptible and Resistant Trees

Plants that are commonly infested by the insects mentioned above are often hosts to sooty molds. Resistance to this condition is normally a result of resistance by the plant to honeydew producing insects.

Control

Control sooty molds by controlling the honeydew producing insect. Chemical control is most commonly used to manage these sucking insects, although biological and cultural control strategies are available for some aphids and scales.One must identify the insect that is causing the problem and the plant being attacked before deciding on a method of management. Consult an appropriate up-to-date OSU Extension fact sheet or bulletin such as Bulletin Perennials). A strong spray of water can be used to dislodge the mold growth from many plants.

POWDERY MILDEWS ON ORNAMENTAL PLANTS

Almost all landscapes have plants that become diseased with one of the powdery mildew fungi. Although the fungi that cause powdery mildew are usually different on different plants, all of the powdery mildew diseases are similar in appearance. In most cases, prompt recognition and control actions can prevent severe damage to plants from powdery mildew diseases.

Symptoms

Powdery mildews, as the name implies, often appear as a superficial white or gray powdery growth of fungus over the surface of leaves, stems, flowers, or fruit of affected plants. These patches may enlarge until they cover the entire leaf on one or both sides. Young foliage and shoots may be particularly susceptible.

Leaf curling and twisting may be noted before the fungus is noticed. Severe powdery mildew infection will result in yellowed leaves, dried and brown leaves, and disfigured shoots and flowers.

Although it usually is not a fatal disease, powdery mildew may hasten plant defoliation and fall dormancy, and the infected plant may become extremely unsightly. On roses, uncontrolled powdery mildew will prevent normal flowering on highly susceptible cultivars.

Hosts

Powdery mildew fungi infect almost all ornamental plants. They are commonly seen only on those plants more naturally susceptible to the disease. Susceptible woody plants include some deciduous azaleas, buckeye, catalpa, cherry, a few of the flowering crabapples, dogwood, English oaks, euonymus, honeysuckle, horse chestnut, lilac, privet, roses, serviceberry, silver maple, sycamore, tulip tree, some viburnums, walnut, willow and wintercreeper.

Powdery mildews are also common on certain herbaceous plants, such as chrysanthemums, dahlias, delphiniums, kalanchoes, phlox, Reiger begonias, snapdragons and zinnias.Remember that each species of powdery mildew has a very limited host range. Infection of one plant type does not necessarily mean that others are threatened.

Environment Favouring Powdery Mildews

Most powdery mildew fungi produce airborne spores and infect plants when temperatures are moderate (60 to 80 degrees F) and will not be present during the hottest days of the summer. Unlike most other fungi that infect plants, powdery mildew fungi do not require free water on the plant surface in order to germinate and infect. Some powdery mildew fungi, especially those on rose, apple, and cherry are favoured by high humidities. Overcrowding and shading will keep plants cool and promote higher humidity. These conditions are highly conducive to powdery mildew development.

Control of Powdery Mildews

Before using fungicides you should attempt to limit powdery mildews by other means. The following cultural practices should be beneficial for controlling powdery mildews.

- Purchase only top-quality, disease-free plants of resistant cultivars and species from a reputable nursery, greenhouse or garden centre. Horticulturists in the green industry and Extension offices should be consulted concerning the availability and performance of resistant varieties.
- Prune out diseased terminals of woody plants, such as rose and crabapple, during the normal pruning period. All dead wood should be removed and destroyed (preferably by burning). Rake up and destroy all dead leaves that might harbour the fungus.
- Maintain plants in a high vigour.
  - Plant properly in well-prepared and well-drained soil where the plants will obtain all-day sun (or a minimum of 6 hours of sunlight daily).
  - Space plants for good air circulation. DO NOT plant highly susceptible plants—such as phlox, rose, and zinnia—in damp, shady locations
  - Do not handle or work among the plants when the foliage is wet.
  - Water thoroughly at weekly intervals during periods of drought.

The soil should be moist 8 to 12 inches deep. Avoid overhead watering and sprinkling the foliage, especially in late afternoon or evening. Use a soil soaker hose or root feeder so the foliage is not wetted.

Chemical Control of Powdery Mildews

In many cases, powdery mildew diseases do little damage to overall plant health, and yearly infections can be ignored if unsightliness is not a major concern.

For example, lilacs can have powdery mildew each year, with little or no apparent effect on plant health. On some plants, powdery mildews can result in significant damage. Thus, fungicides must be used to achieve acceptable control.

ANTHRACNOSE LEAF BLIGHT OF SHADE TREES

A number of different trees are affected by anthracnose diseases. These fungal diseases can cause severe leaf blighting and deformation, but in many cases damage to plant health is not severe. However, with sycamore anthracnose and dogwood anthracnose, the fungus regularly moves back into stem tissue and causes more significant problems. Following are profiles of some of the more common anthracnose diseases of landscape trees.

KABATIELLA APOCRYPTA

Anthracnose diseases are generally not severe on maple, but can cause considerable unsightliness from brownish leaf blotches and some leaf drop when moist weather conditions make the disease particularly severe. Extensive development of stem infections is not common on maples, as it is with sycamore anthracnose and dogwood anthracnose.

The most common symptoms include brownish discolouration along veins, varying from discrete spots to irregular patches of discolouration bordered by veins. Spore masses of the fungus can sometimes be found on lower leaf surfaces along veins during extended moist conditions. The fungus spreads from previously infected tissue in spring to new growth. Where fungicides are used, applications must be started at bud break and continued during early leaf development. When fungicides are required, use labeled products containing mancozeb, thiophanate-methyl, or chlorothalonil.

APIOGNOMONIA ERRABUNDA

This anthracnose disease is primarily a leaf blighting and blotching disease of white ash and, to a lesser extent, green ash. Small twig cankers do occur and the fungus overwinters on these twig lesions, but little damage occurs from this phase of the disease. In wet, cool spring conditions, leaves and sometimes shoots first develop water soaked areas and later large tannish blotches and leaflet distortion.

Considerable leaf drop occurs, especially from lower areas of the canopy. Though this causes concern when leaves litter the ground in late spring, damage to overall plant health is not generally severe and plants typically releaf. As leaves mature they tend to become more resistant to infection. Fungicide applications, if warranted, should be made at bud break, with several repeat applications early in the season. When fungicide is required, use a labeled material containing thiophanate-methyl, chlorothalonil, or mancozeb.

APIOGNOMONIA QUERCINA

White oaks are the most susceptible of many oak species to this leaf blotching disease. Twig infections occur but are not significant except as sources of overwintering fungal inoculum from year to year. Leaves and shoots are infected during cool, wet spring conditions causing leaf blotches that often are strictly delimited by leaf veins.

Eventually, lesions become a papery tan colour and some leaf shriveling occurs. Multiple cycles of infection can occur. Just as leaves near maturity, lesion size lessens, and once leaves mature they become fully resistant by early to mid summer. Fungicides are generally not recommended. If fungicide is required, use a labeled material containing thiophanate-methyl, chlorothalonil or mancozeb

APIOGNOMONIA VENETA

This is a potentially serious disease of American sycamore and to a lesser extent London planetree. Susceptibility of London planetree varies considerably with seed source. Shoot blight, leaf blight and twig and branch cankers and dieback can be severe. The fungus overwinters on twig tissue on the tree with spores splashing to new buds, shoots and leaves in the spring, with disease being enhanced by cool, wet conditions during shoot and leaf development. Considerable defoliation, sometimes with complete leaf loss, occurs on many trees by late spring in some years.

Trees typically releaf by early to mid summer and are less susceptible to continued infections because of warmer, drier conditions. Also, as leaves age they become less susceptible to infection. With repeated infections over the years, cankering of twigs and branches can result in erratic shoot growth that gives an overall distorted appearance to the tree, and also witches- brooming where there are numerous side shoots that develop around a central terminal shoot that was killed by the fungus.

Fungicide applications to prevent infections in the spring are sometimes warranted. If fungicide is required, use a labeled material containing thiophanate-methyl, chlorothalonil or mancozeb. Fungicide injections (which are made by tree care professionals) are also used in spring and fall to systemically control the disease.

DISCULA DESTRUCTIVA

In recent years this disease has become prominent in certain areas, especially on flowering dogwood (Cornus florida). Dogwood anthracnose is most severe where cool, moist conditions occur during the summer, such as in higher elevation areas, and in densely vegetated shady sites with poor air movement. Leaf symptoms include irregular brown blotches bordered in purple on upper leaf surfaces (tan in colour viewed from the leaf underside). Leaf lesions often are delimited by the leaf midvein. Stem symptoms include twig dieback and stem cankering and dieback, often with visible fungal fruiting bodies on dead twigs. Attached wilted, brown leaves often persist into the next spring instead of dropping in the fall. Plants may be killed. Fungal infections occur during moist conditions and fungicides are recommended in the spring from the period of early bud break through bract fall and through early leaf development. When fungicide is required, use a labeled product containing propiconazole or mancozeb. Flowering dogwood should be planted in sites with good soil drainage and adequate organic matter.

Partially shaded sites, such as those with just morning sun are ideal. Plants should receive a moderate fertility programme, and should be mulched to moderate fluctuations in soil temperatures. Dead branches should be promptly pruned from the plant.

Control of Anthracnose Diseases

- Overall tree care programme. Use proper fertilization, pruning, watering, and pest control practices to encourage vigourous plant growth. This aids in general tolerance of the effects of disease and in rapid refoliation in years where disease is severe.
- Fungicide applications. If significant damage occurs yearly and controls are justified, properly applied fungicides may reduce damage from these diseases. High pressure spray equipment will be needed for large tree applications, and this typically requires the hiring of a professional tree care service.

Fungicides will not be effective unless they are applied before and during infection periods. Typically, three applications are necessary, beginning in early spring, with the first application made before leaf buds open. Applications in the fall have been shown to be useful for sycamore anthracnose control. Fungicide injections have also shown promise for sycamore anthracnose control. These must be applied by professional tree care companies.

VERTICILLIUM WILT OF LANDSCAPE TREES AND SHRUBS

Verticillium wilt, caused by the fungi Verticillium albo-atrum and V. dahliae, is a serious vascular disease of hundreds of woody and herbaceous plant hosts. Food crop hosts include everything from raspberries and strawberries, to tomatoes and potatoes.

Some of the many common woody ornamental host plants include ash, barberry, catalpa, elm, magnolia, maple, Russian olive, redbud, smoketree, tuliptree, and viburnum.

One group of plants not susceptible to Verticillium wilt are all the gymnosperms, including conifers such as pine and spruce. While many landscape plants are affected, Verticillium wilt is not a major problem in natural forested areas.

Diagnostic Symptoms

Wilting of leaves and dieback of branches, often one at a time or on one side of the tree, are the most severe symptoms. This can occur over a number of years, with remission of symptoms in some years, or can rapidly progress to plant death in a year or two. Other symptoms of Verticillium wilt may include: marginal browning and scorch of leaves, abnormally large seed crops, small leaves, stunting, poor annual growth, and sparse foliage.

Sometimes large areas of cambial tissue die from infections by the fungus and opportunistic fungi such as Nectria develop in elongated cankers. Late season infections may not be noticeable until plants come out of dormancy with branch dieback evident.

All of the above symptoms can also be caused by other stress factors. A good field symptom that can set Verticillium wilt apart diagnostically is the discolouration of xylem and cambial tissue, visible as streaks if you cut into the wood. This discolouration is variable for different plants: generally greenish to blackish on maple, yellowish green on smoketree, and brown on ash.

This streaking is not totally diagnostic on two counts:

- Other fungi and other factors can cause discolouration; and
- On some hosts and on youngest twigs, infection is not always accompanied by discolouration.

Disease Cycle and Conditions Favouring Disease

The Verticillium fungus can survive for many years in soil, making effective crop rotation difficult. The fungus infects plant roots through wounds and in some cases direct penetration of root tissue. Verticillium also is transmitted from plant to plant by grafting and budding. From root infections, the fungus spreads upward in the plant through the vascular stream.

The results of infection are tissue damage and plugging of xylem, robbing stems and leaves of needed water and minerals. The fungus is returned to the soil as plant parts fall or die, and tiny resistant fungal microsclerotia are spread by wind, in soil and on equipment. Many weed hosts are also susceptible; therefore the cycle of contaminated soil is hard to break. Development of Verticillium wilt is favoured by factors that stress roots, including wounding and droughty conditions.

Control

Disease resistance. If Verticillium wilt is diagnosed at a particular landscape or nursery site it is prudent to replant into that area with a plant that exhibits resistance to this disease. A few common examples of plants typically free of this disease include: crabapple, mountain ash, beech, birch, boxwood, dogwood, sweet gum, hawthorn, holly, katsuratree, honeylocust, oak, pear, London planetree and sycamore, rhododendron, willow, and zelkova. The red maple cultivars Armstrong, Autumn Flame, Bowhall, October Glory, Red Sunset, Scarlet and Schlessinger have also been reported as resistant.

Keep plants as healthy as possible. Proper transplanting practices, proper water management to avoid droughts, a good fertility programme, and pruning out dead branches are all good plant health care management practices. These can help limit infections and help limit the effects of these infections to some extent. Pruning out infected branches is useful as a general horticultural practice for overall plant vigour and aesthetics, but does not eliminate Verticillium from the plant since infections spread from the roots.

Fungicides are not effective for control of this disease. Bacterial Crown Gall on Ornamentals in the Landscape Crown gall is characterized by growth of galls on roots or stems. While mostly found on woody plants, it affects some herbaceous plants as well. Although found on more than 600 plant species in over 90 families, the disease is of economic importance on relatively few ornamental plants. Some commonly affected ornamentals include rose, Prunus, (flowering cherry, flowering almond and ornamental plums), willow, and certain Euonymous species, especially wintercreeper.

Effects of Crown Gall

Crown gall can reduce the productive life of plants. Deformation of tissues due to gall formation disrupts the movement of water and nutrients between roots and leaves. Stems are weakened and growth may be reduced with a general decline in vigour.

The severity of the disease depends on the size, number, and location of the galls, and also on the susceptibility of the plant and age when infected. Galls at the crown of young plants have the greatest adverse effect and can cause stunting and failure to produce healthy leaves and blossoms. This disease may have little noticeable effect on older plants.

Symptoms

Galls may develop anywhere on stems and roots, but are usually found near the soil line. They vary from pea size up to several inches in diameter. Young galls are light coloured and smooth. Older galls become discoloured, hard and woody, and eventually crack, decay and slough off. The texture at first is softer than the normal host stem or root tissue. The galls consist of disorganized host tissues. Secondary galls sometimes form above the sites of the primary gall on stems of some hosts. The secondary galls are usually smaller and occur as separate or unbroken elongated masses of tissue breaking through the bark surfaces. Unlike insect galls, crown galls are a solid mass of tissue all the way through.

Disease Development

Crown gall is caused by the soil-borne bacterium, Agrobacterium tumefaciens. The bacteria can persist in the soil for two or more years even in the absence of susceptible plants. Sometimes the bacteria are carried on seeds. Fresh wounds in stems or roots are essential for the bacteria to invade host tissues.

These wounds commonly occur during planting, cultivation and pruning, and during propagation when grafting and taking cuttings. Soil insects and nematodes can also cause root wounds providing entry sites.

There are no effective chemical controls for this disease in the landscape. Cultural controls include:

- Avoid unnecessary wounding (protect from injury).
- Sanitation-remove infected plants.
- Plant resistant plants in crown gall-infested areas.
- Do not purchase plants with suspicious swelling near the soil line or on the roots.

CEDAR RUST DISEASES OF ORNAMENTAL PLANTS

There are a number of "cedar rust" diseases in which the fungus completes its life cycle on two plant hosts; one in the cypress family and one in the rose family (the rosaceous host). Discussed here are three common cedar rust diseases in the northeast U.S.

- Cedar apple rust (pathogen: Gymnosporangium juniperi- virginianae). The fungus alternates between Eastern red cedar (Juniperus virginiana) and mostly apple and crabapple.
- Cedar hawthorn rust (pathogen: Gymnosporangium globosum). The fungus alternates between junipers and hawthorn, crabapple, and apple in addition to several other rosaceous hosts.
- Cedar quince rust (pathogen: Gymnosporangium clavipes). The fungus alternates between junipers and a wide range of rosaceous hosts. The most noticeable in the landscape is hawthorn.

In some cases these diseases are minor problems, but cedar quince rust and cedar hawthorn rust can be a major problem on hawthorns and cedar apple rust is a major economic consideration in commercial apple production.

Diagnostic Symptoms

Cedar apple rust: On junipers, tan to brownish round to kidney-shaped fungal galls are present in winter and early spring. With moist weather, gaudy bright orange masses of gelatinous spores develop from these galls, and galls swell to several times their original size. Spore masses are several inches in diameter, with a central core and radiating hornlike tendrils, and are highly visible during moist weather in mid-spring.

On apple and crabapple, bright orange-yellow leaf spots develop on upper surfaces of leaves in late spring, followed by light coloured, fringed cup-shaped structures on lower leaf surfaces several weeks later. Damage on junipers is generally minor and involves presence of the galls and twig dieback. On apples and crabapples, fruit infections and leaf drop also can occur.

Cedar hawthorn rust: On junipers, galls are somewhat smaller than with cedar apple rust disease. Galls continue to produce spores on junipers for more than one year, compared to only one season of spore production with cedar apple rust. On hawthorn, leaf spots are similar to above and occasionally green twigs are deformed by the fungus.

Cedar quince rust: Infected areas on juniper are much less spectacular than with cedar apple rust, with a cushion-like mat of orangish fungal growth developing on spherical galls in spring. Cedar quince rust causes the greatest damage of the three rusts to ornamental rosaceous hosts, especially to hawthorns, because of extensive, unsightly fruit infestations, stunting and death of fruits and swelling and distortion of twigs. Infected leaves brown and die. Fruits become covered with orangish-pink spore horns. Unsightly spherical cankers developing on stems can last more than one year.

Disease Cycle and Conditions Favouring Disease

Rust fungi have complicated disease cycles with a number of different spore types that will not be detailed here. A crucial factor relative to control on these cedar rusts, however, is that there is no repeating spore cycle on the rosaceous hosts. In other words, spores produced on hawthorn will not reinfect hawthorns or other rosaceous plants-they will only reinfect junipers later in the season. Spores produced on juniper will not reinfect junipers-they will only infect the rosaceous host. The alternating host plant is necessary for survival of the fungus.

Spores produced on the juniper host are blown during moist weather to the rosaceous hosts in mid-spring at a time when new growth has emerged. The fungus then causes leaf spots on upper leaf surfaces and while growing in the leaf two strains of the fungus mate and emerge as a new spore form on the lower leaf surface.

These spores are then blown back to junipers in mid summer to fall, develop galled areas on the junipers over a one and a half year period and the cycle begins again. Windborne spread of spores between the hosts of several hundred yards is not unusual and spread can be a matter of miles.

Control

Application of fungicides. Protective fungicides can be applied several times starting with prebloom on hawthorn and bud break on crabapples if the disease is chronically a problem at a given site. These applications are to protect the plant from spores being disseminated from the juniper host in mid-spring. Since there is no repeating cycle of this disease on the rosaceous host, further applications after this springtime spread from juniper are unnecessary.

Commonly recommended fungicides include: Mancozeb (Fore, Dithane, Mancozeb); Chlorothalonil (Daconil*); Triadimefon (Bayleton, Strike) and propiconazole (Banner). It is the user's responsibility to follow all label instructions. When you diagnose cedar rust disease from infected hawthorn or crabapple fruits and leaves it is far too late to spray for that year.

Sprays are rarely recommended to protect the juniper host from spores being disseminated from the rosaceous host in late summer and fall.

Eradication of the other host plant. One approach sometimes suggested is to eliminate junipers from around plantings of rosaceous hosts, and vice versa. Concerted efforts to eradicate junipers were historically tried in concentrated apple growing regions.

This practice is limited to some extent by practicality in terms of the widespread occurrence of junipers, long distance spread of the fungi involved, the rights of juniper lovers, and the fact that in most situations cedar diseases are not so serious that such extreme measures are needed. Nevertheless, it is prudent to separate highly susceptible junipers and rosaceous hosts to the extent possible in nursery and landscape situations.

One simple practice where only a few plants are involved is to remove galls from junipers. This is easier to do with cedar apple rust and cedar hawthorn rust, since galled areas are more inconspicuous with cedar quince rust.Use plants with genetic resistance.

A number of juniper species and cultivars and a number of rosaceous plant species and cultivars have varying levels of resistance and susceptibility to these three diseases and where disease pressures are historically high these plants should be used.

TIP BLIGHTS OF JUNIPERS

Phomopsis tip blight and Kabatina tip blight are two common diseases of junipers found in most states east of the Mississippi. Both diseases are caused by fungi and the damage they cause on nursery stock, transplants and certain juniper varieties in the landscape can be severe; however, most established junipers in the landscape are seldom killed. The disease is most serious on younger plants and becomes less serious as plants get older.

There are many varieties of juniper that vary from very susceptible to highly resistant. Junipers are generally considered as low maintenance because they are relatively free of major diseases and insect pests; however, these diseases can adversely affect the appearance and health of these trees in certain locations and under the proper environmental conditions. Although

Phomopsis and Kabatina blights cause almost identical symptoms, aspects of their development and control do differ. Therefore, it is important to distinguish between the two diseases.

Symptoms and Causal Organisms

Phomopsis tip blight, caused by the fungus Phomopsis juniperovora, damages new growth and succulent branch tips of junipers from mid-April through September. Older, mature foliage is resistant to infection; therefore, most blighting occurs on the terminal 4 to 6 inches of the branches.

Affected foliage first turns dull red or brown and finally ash-gray. Small gray lesions often girdle branch tips and cause blighting of foliage beyond the diseased tissue. Small, black, spore-containing fungal fruiting bodies develop in the lesions. Use a hand lens to view these diagnostic fungal structures more easily. Spores of the Phomopsis fungus are produced throughout the summer, and infection can occur whenever young foliage is available and moisture or humidity are high. Most infections usually occur in April through early June and again in late August through September. Very few infections occur in mid-summer or during the winter months.

Repeated blighting in early summer can result in abnormal bunching (witches' broom) and discolouration of the foliage, stunting of young trees or shrubs, or-in severe cases-plant death. Be cautious in diagnosing witches brooming and stunting because similar damage can be caused by the dwarf tip mite. With juniper problems, it is always a good idea to have problem diagnosis confirmed by a diagnostic laboratory.

Kabatina tip blight, caused by the fungus Kabatina juniperi, first appears in February and March; and well before symptoms of Phomopsis tip blight appear. The terminal 2 to 6 inches of diseased branches throughout the juniper first turn dull green, then red or yellow.

Small ash-gray to silver lesions dotted with small, black fruiting bodies of the fungus are visible at the base of the discoloured tissue. The brown, desiccated foliage eventually drops from the tree in late May or June. Foliar blighting occurs only in early spring; it does not continue through the summer. Blighting is also restricted to the branch tips and does not cause extensive branch dieback or tree death. Be cautious in diagnosing witches brooming and stunting symptoms because similar damage can be caused by the dwarf tip mite. With juniper problems it is always a good idea to have problem diagnosis confirmed by a diagnostic laboratory.

The primary infection period for the Kabatina fungus is thought to be in autumn even though visible symptoms are not apparent until late winter or early spring. Infection often is associated with small wounds on branch tips caused by insect feeding or mechanical damage.

Control

- When purchasing new plants, select those that have been reported to have disease resistance. Table has information on relative disease resistance to Phomopsis and Kabatina for several Juniper selections.
- Space new plantings to provide good ventilation and air circulation and avoid heavily shaded areas. Avoid wounding plants, especially in spring and autumn.
- Water plants in early morning so the foliage will dry as soon as possible. Maintain adequate fertility, but do not over fertilize.
- Prune out diseased branch tips during dry summer weather and destroy them. Do so only when plants are dry and no rain or overhead irrigation is expected for several days. Avoid excessive pruning or shearing.
- Chemical control of these tip blight diseases normally is not necessary in established landscape or windbreak plantings. Occasionally, fungicide applications may be needed on susceptible junipers to control Phomopsis blight. Application of certain copper- based fungicides (Phyton-27, Kocide), thiophanate-methyl (Cleary's 3336, Domain, Fungo FLO), or mancozeb (Fore, Dithane, mancozeb) at 7- to 21-day intervals during rapid plant growth in the spring will give adequate control of Phomopsis but not Kabatina tip blight. Kabatina blight infections occur in the fall, and there currently are no fungicides labeled for control of this disease.

RHIZOSPHAERA NEEDLECAST ON SPRUCE

Spruce, in particular, Colorado blue spruce, can be infected with a needlecast disease caused by the fungus Rhizosphaera kalkhoffii. Trees planted in nurseries, Christmas tree plantations, and landscapes can be infected. Trees are not usually killed by this disease; however, premature needlecast results in trees that are not marketable, or which are not acceptable in the landscape.

Symptoms and Disease Cycle

A healthy spruce will retain its needles 5 to 7 years. A spruce severely infected with Rhizosphaera needlecast may hold only the current year's needles. Rhizosphaera needlecast infects needles on the lower branches first and gradually progresses up the tree. This pattern holds true for most needle diseases on conifers and is the result of more favourable conditions for disease development near the ground. Under epidemic conditions, lower branches may be killed by this fungus. Although needles on new growth become infected in May and June, symptoms are not visible until late fall or the following spring, when infected needles turn purple to brown and begin to drop. Tiny fruiting bodies of the Rhizosphaera fungus protrude through the stomata of the infected needles. Under a hand lens, these stomata appear as fuzzy black spots instead of their usual healthy white colour. During wet weather in late spring, spores are released from these fruiting bodies and are rain splashed onto newly developing needles where infection occurs and the disease cycle is repeated.

Cultural Control

Very little is known about cultural control of Rhizosphaera needlecast. The following guidelines will help prevent serious losses.

Use Healthy Stock

At planting time, the foliage of blue spruce should be examined for fruiting bodies of Rhizosphaera protruding through needle stomata. If these bodies are present, the tree should not be planted.

Maintain Tree Vigour

Although detailed studies are lacking, it has been observed that trees suffering from environmental stresses are often more seriously attacked by Rhizosphaera. Spruce are particularly sensitive to heavy, compacted soils which become quite dry in late summer. Vertical mulching such soils to improve aeration and water penetration may help lessen the severity of the disease. Root irrigation during dry weather should also be carried out whenever possible.

Prevent Spread by Shearing Tools

Shearing when the foliage is wet may result in spread of the spores on shearing tools. To avoid this possibility, do not shear infected trees when the foliage is wet (such as when dew is on the foliage in the morning).

Shear healthy trees first to avoid carrying the spores from a diseased tree to a healthy one. If this is not possible, tools should be sterilized after shearing a diseased planting. Denatured alcohol, available at most paint stores, will kill the spores and also remove pitch from tools. A three- to five-minute dip will do the job.

❒

Index

L

M

N

P

R

S

V

W

X

❐❐❐